Lecture Notes
in Control and Information Sciences 215

Editor: M. Thoma

Springer-Verlag London Ltd.

Claudio Bonivento, Giovanni Marro and
Roberto Zanasi (Eds)

Colloquium on Automatic Control

 Springer

Editors

Professor Claudio Bonivento
Professor Giovanni Marro
Dr Roberto Zanasi
Dipartimento di Elettronica Informatica e Sistemistica
DEIS, Università di Bologna
Viale Risorgimento, 2, 40136 Bologna, ITALIA

ISBN 978-3-540-76060-3   ISBN 978-3-540-40948-9 (eBook)
DOI 10.1007/978-3-540-40948-9

British Library Cataloguing in Publication Data
Colloquium on automatic control. - (Lecture notes in
  control and information sciences ; 215)
  1.Automatic control
  I.Bonivento, C. II.Marro, Giovani III.Zanasi, Roberto
  629.8

Library of Congress Cataloging-in-Publication Data
A catalog record for this book is available from the Library of Congress

Typesetting: Camera ready by editors
69/3830-543210 Printed on acid-free paper

Preface

This volume collects the text of the invited lectures purposely prepared for the Colloquium on Automatic Control which was held at the University of Bologna on June 10th, 1996. Offering a complete picture of all the scientific interests and technical implications characterizing the wide and deep research area of systems and control was beyond the scope of this event. For this, the reader could rather examine the really impressive technical programmes of the principal conferences of the field, such as the IFAC Congress to be opened in San Francisco at the end of this month or the AACC series. More simply, the initiative of the Colloquium and the present publication moved firstly from the desire to underline the half century since the pioneering start of the research activity at the University of Bologna in the control area, which can be identified with the appearance of the book entitled *The Regulation of Hydraulic Turbines* by Giuseppe Evangelisti in 1947. Secondly, this local anniversary would have given us the special opportunity of focusing in an international setting a fruitful discussion on selected research themes, which among others are nowadays central for the development of Automatic Control as a science and a technology in service of civil and industrial society.

We are deeply grateful to all the colleagues and friends from various other Universities who accepted our invitation with enthusiasm and have contributed so significantly to the high scientific quality of the Colloquium. In particular, we wish to thank Karl Johan Åström who offers a brilliant and stimulating opening lecture on *Automatic Control: A Perspective* where he shows how the science of control emerged from attempts to solve problems in such diverse areas like ship steering, telecommunication, process control and flight control. He then discusses motivations, ideas and people relevant to the establishment in the last few decades of the main approaches to control, ending with a scenario of the factual content of the field at present and some personal reflections on its future. The Sigillum Magnum of the University of Bologna awarded to K.J. Åström on this occasion is not only a further acknowledgment for his outstanding scientific contributions but at the same time a sign of sympathy and friendship from our scientific community.

Alberto Isidori in the second lecture on *Semiglobal Robust Regulation of Nonlinear Systems* illustrates a new unifying approach to the problem of output regulation of nonlinear systems which yields necessary and sufficient conditions for local structurally stable regulation and, for special classes of nonlinear systems, also semiglobal robust regulation.

Edoardo Mosca in the third lecture on *Nonlinear Predictive Command Governors for Constrained Tracking* presents a method which is based on conceptual tools of predictive control and solves tracking problems in presence of pointwise-in-time input and state inequality constraints by means of the design of a nonlinear device devoted to the generation of virtual input and state reference sequences.

Giovanni Marro in his *Multivariable Regulation in Geometrie Terms: Old and New Results* offers a critical review of the basic tools of the geometric approach presented by several authors and by himself in particular over a period of more than 25 years, along with the solutions of the central issues of regulation, that is, asymptotic tracking, disturbance rejection and, when possible, perfect tracking.

Claudio Maffezzoni and Gianantonio Magnani give the lecture on *Practice and Trends in Control Engineering* analysing the current state of control engineering methods available for industrial applications, the principal needs demanding more systematic and reliable design methods and the lines along which engineering research moves to meet such demands.

Claudio Bonivento and Roberto Zanasi in the last lecture on *Advances in Variable Structure Control* address the problem of chattering reduction which is crucial for the practical application of the sliding-mode theory. The Discontinuous Integral Control approach proposed by the authors is reviewed both in the continuous and discrete-time context with a discussion of some real-world application cases.

In our opinion, the book contains a number of well-balanced theoretical and engineering-oriented elements usable by different classes of readers, ranging from non specialists to experts and professionals in the

field. For all of them we hope this effort may represent a significant
contribution for cultural and technical progress.

Finally, we acknowledge with pleasure the active collaboration with
Springer-Verlag London and in particular with Christopher Greenwell
for his competent assistance during the editorial work.

Claudio Bonivento
Giovanni Marro
Roberto Zanasi

Bologna, June 1996.

Contents

1. Automatic Control: A Perspective

Karl Johan Åström*

Abstract

This paper presents some personal reflections on the field of automatic control, its emergence, development, and its future. The assessment is that automatic control has been very successful, but that we now are at a crossroad, where we have to decide if we want to take an holistic view with full systems responsibility or if we want to fractionate into a collection of subspecialities.

1 Introduction

Research and education in engineering were originally focused on specific technologies such as mining, machines, electricity, and industrial use of chemistry. This led to organization of engineering schools with departments of mining, mechanical engineering, electrical engineering, and chemical engineering etc. This served very well in the beginning of this century. The situation changed significantly with the advent of fields like automatic control, that cut cross traditional department boundaries.

Automatic control is crucial for generation and transmission of energy, process control, manufacturing, communication, transportation and entertainment. It has been a key element in the design of experimental equipment and instrumentation used in basic sciences. Principles of

---

*Department of Automatic Control, Lund Institute of Technology, Lund, Sweden.

automatic control also have impact on such diverse fields as economics, biology, and medicine.

Automatic control, like many other branches of engineering science, has developed in the same pattern as natural science. Although there are strong similarities between natural science and engineering science, it is important to realize that there are some fundamental differences. The inspiration for natural science is to understand phenomena in nature. This has led to a strong emphasis on analysis and isolation of specific phenomena, an extreme case being particle physics. A key goal is to find basic laws that describe nature. The inspiration of engineering science is to understand, invent, and design man-made technical systems. This places much more emphasis on design. It is natural to consider interaction instead of isolation. A key goal is to find system principles that makes it possible to effectively deal with complex systems. Feedback, which is at the heart of automatic control, is an example of such a principle.

The wide applicability of automatic control has many advantages, but it has also created some difficulties. Because automatic control can be used in so many different fields, it is a good vehicle for technology transfer. A key difficulty is, however, to position the field academically. One approach is to have separate control groups within traditional departments of electrical, mechanical, and chemical engineering, another is to have a central group with global responsibility and strong connections to applications. One advantage with the centralized approach is the ease ideas can be transferred among different fields. A disadvantage is the difficulty to maintain expertise in different application areas, which is indispensable for successful use of automatic control.

In summary, automatic feedback control is a young exciting field. It has found widespread application in the modern technological society, many systems cannot be built without feedback. There is a dynamic development with many challenging problems remaining to be solved.

2   The Roots

Although we can find very early examples of the use of feedback, the development of automatic control is strongly connected to the industrial developments associated with the industrial revolution. Whenever new sources of power were discovered the need to control them immediately arose. When new production techniques were developed there were needs to keep them operating smoothly with high quality. In this section we will follow some of the roots of automatic control. We will start with the consequences of the development of steam power and industrial production. We will then discuss some implications regarding the development of ships and airplanes and we will end with some consequences for the emergence of telecommunications.

2.1 Governors

The need for control devices appeared already in the operation of windmills. The famous centrifugal governor first appeared there. The first major application, however, was in connection with the development of steam power. The desire to keep textile mills running at constant speed was a strong motivation. The result was an interesting development of governors for machines over a 100 year period stretching from late 1700. Theoretical understanding of the devices also developed starting with the papers [39] and [58]. They provided basic understanding and posed interesting stability problems, which were later solved independently by [49] and [23]. There was little interaction between the scientists. Routh and Hurwitz were not aware of each others contributions and neither knew about the work by Lyapunov [38]. Vyshnegradskii's paper had a very strong influence on engineering practice and was widely used in control systems for hydroelectric power. The book [53] remained standard material for control of machines for a long time. This book gave a good presentation of the state of the art of governors at the turn of the century. The idea of feedback and the standard PID controller are well exposed. Methods of analysis based on linearization and analysis of the roots of the characteristic equation based on Hurwitz theorem are also well presented.

2.2 Industrial Process Control

Automation of process and manufacturing industries evolved from the late 1800. The development of control in that field was largely parallel to the development of governors. It resulted in the standard PID controller now commonly used in industry. Proportional feedback as well as integral and derivative action was rediscovered because the workers in the field failed to see the similarity with the already existing work on governors. There were very modest theoretical developments. The state of the art is illustrated by the following quote from [24]: "In spite of the wide and ever-increasing application of automatic supervision in engineering, the science of automatic regulation of temperature is at present in the anomalous position of having erected a vast practical edifice on negligible theoretical foundation." One explanation is that analysis was difficult because of the fact that most process control systems had time delays, another is that the industry structure was such that it permitted few luxuries and that hardware problems were still predominant.

One lasting contribution is an empirical method for tuning the standard PID controller [61]. It is interesting to observe that this work made extensive use of simulation of closed loop dynamics using the differential analyzer which had just been developed by Vannevar Bush, see [15].

2.3   Ship Steering

There were many interesting developments in ship steering. The word servo motor was coined by the French engineer Farcot who developed hydraulic steering engines.

Autopilots for ships were developed based on gyros and magnetic compasses. Sperry was one of the most successful engineers who dominated the commercial field with more than 400 systems installed in 1932. Sperry's design was largely intuitive where he tried to mimic the behavior of an experienced helmsman. By observing experienced pilots Sperry came to the conclusion that "An experienced helmsman should 'meet' the helm, that is, back off the helm and put it over the other way to prevent the angular momentum of the ship carrying it past the desired heading." This effect can be captured very well by using derivative action. Sperry's succeeded very well and his autopilot, which was nicknamed the Metal Mike was very successful because it captured much of the behavior of a skilled pilot.

There was also interesting theoretical developments. The paper [40] gives a taxonomy of a wide class of controllers which leads to recommendation of a controller of the PID type. The results were also verified in full scale experiments. In spite of this Minorsky's work had little influence on the actual design of autopilots for ships.

2.4   Flight Control

There were many experiments with manned flight in the end of last century. The Wright brothers were the first to be successful. One major reason was that they clearly understood the relation between dynamics and control. This was expressed very clearly by Wilbur Wright in 1901. "Men already know how to construct wings or airplanes, which when driven through the air at sufficient speed, will not only sustain the weight of the wings themselves, but also that of the engine, and of the engineer as well. Men also know how to build engines and screws of sufficient lightness and power to drive these planes at sustaining speed ... Inability to balance an steer still confronts students of the flying problem. ... When this one feature has been worked out, the age of flying will have arrived, for all other difficulties are of minor importance."

By using their insight and skillful experiments the Wright brothers made the first successful flight with Kitty Hawk in 1905. The fact that this plane was unstable was a strong impetus for the development of autopilots based on feedback. Sperry made a very spectacular demonstration of his autopilot as early as 1912. His son was flying a plane close to the ground with his hands in the air and the mechanic was walking on the wing! Interesting work on autopilots were also made in England and Germany in the same period. It is also interesting to note that a fully automatic flight across the Atlantic was made by the Robert E. Lee on September 23, 1947.

Although a good theoretical understanding of flight dynamics was based on linearized equations and analysis of the characteristic equation based on Routh's stability criterion was available as early as 1911, the theoretical work did not have any impact on practical autopilot design until the mid-1950s. One possible reason is that tools for computation were lacking.

2.5 Telecommunication

Feedback was an enabling factor in the development of telecommunication. The key contribution was the invention of the feedback amplifier by Black. In his efforts to obtain a highly linear amplifier that was very insensitive to variations in the characteristics of the electron tubes Black had made the observation that: "... by building an amplifier whose gain is deliberately made, say 40 decibels higher than necessary and then feeding the output back on the input in such a way as to throw away the excess gain, it had been found possible to effect extraordinary improvement in constancy of amplification and freedom from non-linearity."

The importance of Blacks invention is illustrated by the following quote by Mervin B. Kelley at the presentation of the IEEE Lamme Medal in 1957. "Although many of Harold's inventions have made great impact,

that of the negative feedback amplifier is indeed the most outstanding.
It easily ranks coordinate with De Forest's invention of the audion as
one of the two inventions of broadest scope and significance in electronics
and communications of the past 50 years ... it is no exaggeration to say
that without Black's invention, the present long-distance telephone and
television networks which cover our entire country and the transoceanic
telephone cables would not exist."

An interesting fact about Blacks invention was that nine years was
used in the patent office because the officers refused for a long time to
believe that the amplifier would work.

Blacks work also inspired much theoretical work, for example the
stability analysis [42] and [12] which all became pillars of automatic
control.

3   The Field Emerges

Around 1940 there were a large number of control systems based on feedback in operation in a wide range of fields as discussed in the previous
section. Although some ideas were the same, for example linearization
and analysis of the closed loop characteristic equation, the commonality
of the systems were not appreciated. One of the most striking illustra-
tions of this was that feedback was reinvented too many times. Another
is that the nice properties of integral and derivative action was also
rediscovered.

By the late 1930s the attractive properties of feedback were reason-
ably well understood in many disciplines. For example that feedback
could give systems with highly linear responses that were insensitive to
disturbances and parameter variations. This is expressed very well in
the seminal papers [22] and [21]. "For a simple vacuum tube amplifier
the linear response was due to to the constancy of the parameters within
the amplifier. Any departure from constancy of these parameters affects
the relation between input and output directly. For a servomechanism,
however, the only function of the servo-amplifier element is to apply sufficient force to the servo output to bring it rapidly to the correspondence
of the servo input. Such an amplifier can be a relatively crude affair."

In Blacks patent on feedback there are also claims that feedback can
be applied not only to electronic amplifiers but to a wide range of other
systems, see [10] and [11]. Similar ideas are expressed very clearly in
the following quote from [45]. "But there was no common approach to

control problems in the different engineering disciplines. For example in 1937 I gave a conference paper on flight control, which was later published in an aeronautical journal. This was the first systematic study of the topic, and at the end of the paper I also compared flight control with the control of other variables such as pressure and temperature, giving the equivalent mechanical system in each case and comparing systems equations and typical dynamic behavior. This idea of a common ground between entirely different types of control systems was quite new but the paper was not known outside aeronautics."

3.1 War Pressures

Although there were glimpses of a pattern there was still not enough structure to call it a discipline. Although there were techniques to analyze the stability of a system the existing methods did not indicate how the system should be modified to be stable. Neither were there any systematic methods to design a control system with specified properties.

It is a reasonable statement that the field of automatic control emerged as a discipline during the Second World War when it was realized that science could have a drastic impact on the war effort. The development was strongly stimulated by the cross disciplinary teams that were formed. Similar things happened in many countries, most notably in the US where good documentation is also available. The strongest driving force came from the development of radar and fire control. In the US there was a National Defense Research Committee (NDRC) directly under President Roosevelt. This committee organized and funded the Radiation Laboratory at MIT. At one time this laboratory had about 4000 researchers, most of them from outside MIT. Fire Control was one of the themes where much work on automatic control was done. A characteristic feature of the laboratory was a strong interdisciplinary staff with a wide range of backgrounds. A major contribution of the Radiation laboratory was the development of systematic design methods based on the techniques that were developed at Bell labs by Bode and Nyquist.

Major contributions were also given by other Laboratories at MIT notably the Instrumentation Laboratory and the Servomechanisms Laboratory. This laboratory, which had a very dynamic leader Professor Gordon S. Brown, was based on the research on servomechanisms done by Hazen.

3.2   Servomechanism Theory

Since so many applications centered around servo problems the resulting theory became known as servomechanism theory although there were major applications in a wide range of fields. The Radiation Laboratory was dissolved immediately after the war. The research work at the laboratory was published in an impressive series of 28 volumes. The book [25] dealt with the theory of servomechanisms. The book [13] by authors from the Servomechanism Laboratory followed shortly after. To get a perspective on the development it is also enlightening to read some of the original reports. This also gives a clear indication that there was a very wide network of researchers working on the problems.

The servomechanism theory contained a graphical representation in terms of block diagrams that is independent of the technological details of the controlled system, a mathematical theory based on linearization, complex variables, and Laplace transforms. The transfer function of a system could be determined experimentally by investigating the propagation of sine waves through the system. In this way it was possible to deal with systems where physical modeling was difficult. A systematic method for designing a controller that gave a closed loop system with specified properties was a key element. The particular method developed was based on graphical methods,using Bode and Nichols diagrams.

The servomechanisms theory was used extensively in the industries involved in the war time efforts and it spread rapidly. The availability of good books contributed to this. Among the early books we can mention [43], [55], [46], [17] and [54]. In view of the wide application range it is perhaps more adequate to call the approach *frequency response*. This captures the idea that much of the analysis and design focused on the notion of frequency response. The name is also not so strongly tied to a specific application.

The book by Tsien called Engineering Cybernetics was very advanced for its time. It expressed the fact that something new had really happened in the following way: "A distinguishing feature of this new science is the total absence of considerations of energy, heat, and efficiency, which are so important in other natural sciences. In fact, the primary concern of cybernetics is on the qualitative aspects of the interrelations among the various components of a system and the synthetic behavior of the complete mechanisms." Later he goes on to say: "An engineering science aims to organize the design principles used in practice into a discipline and thus to exhibit the similarities between different areas of engineering practice and to emphasize the power of fundamental concepts." Another very interesting feature of the book is that it anticipated much of the future development.

The book [54] filled in an important gap in implementation because it gave a systematic way to design a controller using analog techniques.

3.3   Education

Automatic control was recognized as a very powerful technique that could be applied to many problems in diverse fields. It quickly received world wide acceptance in industry and academia. Control groups were created in many companies, and new industrial enterprises specializing in control were established. Courses in the field were introduced at practically all engineering schools.

3.4   The Second Wave

Automatic control in the form of servomechanism theory or frequency response was very well established in the beginning of the 1960's with a wide range of applications. One factor that strongly contributed to the excitement of the field was that shortly after its establishment there was a second wave. There was a substantial injection of ideas from several different sources. The space race which began with Sputnik in 1957 was a strong driving force. Computers started to be used for implementation of control systems, and there was an influx of ideas from mathematics.

The development of space travel posed many challenges. There were many optimization problems, for example: How to make the most effective use of moderate sized rocket to put the largest payload in orbit? How to find efficient trajectories for interplanetary travel? How to minimize heating at reentry to the earth's atmosphere? Attempts to solve problems of this type led to the development of optimal control theory. Major contributions were made by mathematicians and control engineers. The development led to a strong revitalization of the classical calculus of variations. Pontryagin and his coworkers in Moscow followed the tradition of Euler and Lagrange when they developed the maximum principle, see [47]. Bellman instead followed the ideas of Hamilton and Jacobi when he developed dynamic programming, see [7], [8]. The special case of linear system with quadratic criteria was solved by Bellman in special cases and in [28]. The books [6] [14] presented the results in a form that was easily accessible to engineers. These book also dealt with computational issues.

The idea to control industrial processes using digital computers emerged in the mid 1950s. Louisiana Power and Light Company installed a Daystron computer for plant monitoring in September of 1958 and Texaco installed a RW-300 at their Port Arthur refinery. This system did closed loop control for the first time on March 15, 1959. The use of digital computers then expanded very rapidly and computer control is today the standard technique for controlling industrial processes. The theory of sampled system had already started to be developed in connection with air defense systems. Early work was done at Columbia University in the US, in the UK and at the Institute of Control Problems in Moscow. The early contributions are well described by the key contributors in their textbooks [48], [27] and [56].

Disturbances is a key ingredient in a control problem, without disturbances and process uncertainties there is no need for feedback. Modeling of disturbances is therefore important. In [25] and [51] it was proposed to use stochastic processes as models for disturbances. A key problem in fire control was to predict the future motion of an aircraft. Very elegant solutions to this problem were given independently by Wiener and Kolmogorov, see [59] and [34]. A reformulation by Kalman and Bucy led to the Kalman filter as a solution which is very attractive computationally, see [31] and [29].

Stochastic control theory emerged from a combination of optimal control with the theory of stochastic processes. In the case of linear systems with quadratic criteria it led to a very elegant separation of the complete problem into a combination of a Kalman filter and a state feedback from the estimated states, see [1]. This separation does not hold for general systems. The solution then becomes very complicated. It leads to dual controllers where the controller has the dual task of driving the system towards its goal but also to inject signals so that we obtain better information about the state of the system, see [19].

The theoretical approaches to control were originally based on differential equations. The servomechanism theory was a radical change because the systems were now characterized by their frequency responses. The influx of ideas during the 1960s led to a return to differential equation models which at the time was called state space theory, see [60]. Many fundamental questions were also raised such as controllability and observability, [33]. Algebra was also used heavily to develop the theory, see [30] and [32].

Frequency response was a very powerful method to determine a transfer function experimentally. With the emergence of state-space theory came a need for experimental methods to determine appropriate systems models. This coincided with the availability of digital computers. The result was a dynamic development of the field of system identification, see [2].

Initially the emergence of all new ideas led to quite a bit of controversy between "classical" and "modern" control but the viewpoints have now largely been unified and most professionals agree that it is very valuable to have several views of a control problem. The development of both approaches led to a very dynamic and rapid development of the field.

3.5 Organization

Much of the early work in automatic control was classified because of its military connection. Many researchers felt that there were strong benefits from an open exchange of ideas, which resulted in a strong drive for internationalization. This need became very strong after the war. In 1956 there were plans for no less than eight national meetings on automatic control in Europe. Good leadership resulted in the formation of an international body IFAC (International Federation of Automatic Control), which became the international forum for our field, see [44], [16] and [37]. An organizational structure was set up, tri-annual World Congresses, symposia, and workshops were organized, and a journal was started. A key meeting in IFAC is the tri-annual World Congress with the first World Congress held in Moscow in 1960. In preparation for this a meeting was held in Cranfield, UK, see [57]. This was probably the first manifestation of Automatic Control as an established field. The activities have expanded substantially and today there are IFAC meeting almost every week.

Although IFAC is the global international organization for automatic control there are also other important actors. The Automatic Control Council in USA, which is the national member organization of IFAC in USA, organizes the yearly American Control Conference in collaboration with many engineering societies. The European Control Conference, which meets every other year, initiated with a meeting in Grenoble in 1991. An Asian Control Conference had its first meeting in 1994.

Automatic control also has an important place in the meetings of many engineering societies, AIAA, AIChE, AISE, ASME, IEEE, ISA, and SCS in cooperation with IFAC.

The establishment of journals is an important aspect of a scientific field. A number of high quality journals appeared simultaneously with the emergence of organizations.

3.6   An Italian Perspective

So far we have give a global view. An interesting perspective is also obtained by viewing the development from an individual country. On this occasion it is particularly appropriate to have an Italian perspective. Professor Evangelisti in Bologna was one of the true pioneers of Automatic Control in Italy. He graduated in Civil Engineering and became full professor at the Institute of Hydraulic Construction in 1939. He devoted considerable time and energy on research on regulation problems. His book [18] is one of the first books on control problems in hydraulics. In 1957 Professor Evangelisti founded the Servomechanism and Computing Center at the University of Bologna. It is interesting that already at that time he saw the advantage in combining control and computation. Professor Evangelisti also signed the original resolution that led to the formation of IFAC in 1956. Education in automatic control started in several other universities in the late fifties. In 1961 three researchers, Antonio Ruberti, Antonio Lepschy and Giorgio Quazza were given the title "libero docente" in Automatic Control. Quazza had graduated from Polytechnic of Turin and studied at MIT and at the Brooklyn Polytechnic Institute. He gave seminal contributions to control of power systems and he was very active in IFAC. His name will always be remembered in the control community because the highest technical award in IFAC bears his name. Ruberti was very influential for the control community in Italy by establishing a series of meetings on Control and System Theory which has been running annually since 1963. He also promoted close scientific relations with scholars in the US. First full professorships were established in the middle of sixties, Professor Ruberti in Rome, Professor Belardinelli in Bologna, Professor Lepschy in Padua, Professor Biondi in Milan, Professor Volta in Genoa and Professor Piglione in Turin. The wisdom and vision of these pioneers was instrumental for a very fruitful development of automatic control in Italy.

4 Current Status

One way to find the factual content of a field and how it has developed is to investigate the contents of the leading conferences in the field over a number of years. A perspective on the development can be obtained by analyzing the contents of the IFAC World Congresses, the American Control Conference, and the IEEE Decision and Control Conference over the period 1960–1995. Complementary views are given in position papers, e.g., [35].

4.1 System Theory

In practically all meetings in the period there are sessions on linear systems (time invariant as well as time-varying), nonlinear systems, stochastic systems, discrete time systems, time delay systems, distributed parameter systems, and decentralized systems. Occasionally there has also been interest in special topics such as variable structure systems. These sessions naturally have a strong flavor of applied mathematics. The issues discussed are representations, development of concepts such as stability, observability, controllability, analysis of structure etc. The tools used are typically ordinary and partial differential equations, but many other mathematical tools have also been explored. There is, e.g., a significant interest in exploring the relations between differential geometry and nonlinear systems. Lately there has been a significant interest also in differential algebraic systems, sometimes under the label of "behavioral systems" and in discrete event dynamical systems. The field of linear systems has been declared uninteresting at many instances but interest has often been renewed due to new viewpoints.

4.2 Modeling and Identification

Models are a very important element of automatic control. Modeling is a significant portion of the development of all control systems. It is also an area where specific process knowledge plays a major role. There are two approaches to modeling, one is based on physics (white-box modeling) the other on on plant experiments (black-box modeling). It is also possible to combine the approaches. This has recently been popularized under the name of grey-box identification. The early effort was focused on experimental techniques based on frequency response. There has been a significant effort in system identification. This is a very fertile field with connections to many branches of applied mathematics. Lately there has been a focusing of the interest because of a significant interest in identification for control. The area of model reduction is a closely related field, which comes up regularly in the meetings.

4.3 Design

It is difficult to capture design problems formally because there are so many factors to consider. A common approach has been to formalize one or a few aspects of the problem so that it can be dealt with mathematically. A lot of creativity has been demonstrated in this area. The early work had a strong engineering flavor with emphasis on graphical methods like loop shaping and root locus. Analytical design methods such as pole placement were introduced later. Optimization is a natural tool that has stimulated much research on optimal control. It comes in many flavors, LQG, LTR, dynamic programming, and the maximum principle. Because of the nature of the problem and the caliber of the people involved much interesting mathematics has also appeared. A focus on disturbances led to the development of stochastic control theory. In the last decade there has been a significant effort to consider model uncertainty. This resulted in robust control techniques like QFT, H_∞, L_1-methods, μ, and LMI. A gain there has lately been a significant interest in nonlinear methods such as feedback linearization.

Finding fundamental limitations for achievable performance is a basic aspect of control system design. The early work by Bode gave considerable insight into this problem, but it is an important area that deserves much more research. It is particularly important for the growing area of integrated control and process design.

Control system design is a difficult area. If all engineering aspects are considered the problem becomes very messy. If the problem is simplified so that it can be solved elegantly many aspects have to be neglected. This has caused much controversy. It will probably take a long time before we have good design tools that will deal with the real issues of control system design. An important requirement is that the methods also should be developed and packaged so that they can be used by an engineer with a training at the master's level.

An indication of the importance to have techniques that can be used by persons with a moderate technical background is that special design methods have appeared in application areas with very little initial contact with the main stream control theory. Typical examples are dynamic matrix control (DMC), in chemical process control, and the computed torque method in robotics. A dynamic matrix is nothing but a truncated impulse response for a multi-variable system. Since the chemical engineers were ignorant of this, they solved the problem and invented a new vocabulary (DMC, QDMC, IDCOM). There is a good lesson to be learned from this. We must package our results so that they are accessible. The following quotation from a speech by Louis V Gerstner, Jr., (the chairman of IBM) gives a useful perspective: "The information revolution will happen, but only when the industry stops worshiping technology for its own sake and starts focusing on *real value* for the customers." Even as researchers I believe that we must occasionally stop to think about our customers.

4.4 Learning and Adaptation

To have systems that could automatically learn about a process and do control design automatically has been a long standing dream among control engineers. There have been sessions on this from the 1960 Moscow Conference and onwards. This has led to the model reference adaptive controller and the self-tuning regulator. It is, however, only lately in the 1980s that the simpler problem of providing control systems with devices for automatic tuning of the controller has been approached. This has had a significant impact particularly on the field of process control. Much work, however, remains before the longstanding goal is fulfilled. There are strong connections to the problem of machine learning—an area that is currently developing in a very interesting way.

4.5 Computing and Simulation

The development of computational tools has been very important for the development of automatic control. Computations are used when analyzing systems, and they are also an integral part of a controller. Much of the early development was hampered by poor equipment. This is very noticeable in the development of autopilots and industrial PID controllers. The following quote by Vannevar Bush from 1927 gives an interesting perspective. "Engineering can proceed no faster than the mathematical analysis on which it is based. Formal mathematics is frequently inadequate for numerous problems pressing for solution, and in the absence of radically new mathematics, a mechanical solution offers the most promising and powerful attack ... ." The mechanical differential analyzer developed by Bush 1928-1931 was a very influential contribution and there was more to come.

Automatic control has been strongly linked to computing throughout its history. Computing is important for analysis, simulation, and implementation. Analog computing was used initially but has lately been replaced by digital computing. The development of the computer is probably the factor that has changed the most in the short span of the field. An illustration of this is that at the First IFAC World Congress in Moscow computing appeared under the sub-heading "Pneumatic Components and Computing Devices" in the components category! The emergence of computers have also put much more emphasis on numerics, which is important both for analysis and implementation. Lately there have been both special sessions and special symposia on Computer Aided Control Engineering. The emergence of interactive software like MATLAB and Matrix$_X$ has also had a significant influence on the field. In later conferences we can find several special sessions on this. Simulation is a very important area that now has its own conferences and publications. It is also interesting to observe that there are numerical analysts that have taken a real interest in problems that arise from analysis and design of control systems.

4.6   Implementation

Implementation is an important issue. Because of the rapid development of micro electronics it is also one of the factors that has changed the most over the past 50 years. At the Moscow Congress there were three major groups: theory, components, and applications. Implementation issues came under components. Some aspects on implementation, such as sampled data systems, are very well represented but many equally important aspects are not dealt with at all. There has, e.g., only been a few sessions on windup over the years. Many important aspects on implementation are not covered in textbooks. Typical examples are windup, real time kernels, and man machine interfaces. In my mind this is probably one of the factors that has contributed most to the GAP between theory and practice.

4.7   Commissioning and Operation

There are a wide range of interesting problems that appear after a control system is designed and implemented. First the system has to be commissioned. Then it is necessary to continuously supervise and assert that the system is running properly. Many of the problems occurring during this phase have only recently been addressed. Typical issues are fault detection and diagnosis, but there are also many other interesting problems, such as loop assessment and performance assessment. Development in this area is strongly motivated by the drive for higher quality.

4.8 Education

Education has been well represented at the conferences. The sessions have naturally been dominated by academics. Lately there has been efforts to bring in industrialists.

4.9 Applications

The broad application areas have not changed much over the years. At the Moscow Conference there were sessions on metalworking, electrical power systems, electrical drives, transport, industrial processes, chemical and oil, thermal and nuclear power, and metallurgical processes. In later years we can also find applications in aerospace, automotive, micro electronics, and consumer electronics (CD, Video etc). A long experience with journals and conferences, however, has shown that it is very difficult to get good applications papers. The engineers who really know about the applications do not have the time or the permission to publish. Many of those who do write have only a superficial knowledge about the applications. This sends distorted signals in all directions. There are occasional efforts with special issues of journals, where really good applications papers sometimes appear. We need those badly for better education of the next generation of control engineers.

5 The Future

Automatic control has emerged as an interesting and successful discipline in a period of about 50 years time. The development of the field has been very dynamic. It is one of the first examples of a systems discipline that transcends the traditional boundaries of the traditional engineering disciplines. Automatic control has a firm base in applied mathematics and an unusually wide range of applications. Automatic control is "mission critical" for many things in modern life, for example, in airplanes and CD-players. A natural question to ask is if the development will continue or if it has reached its apex. Having a positive attitude I strongly believe in a continuation, but what happens will of course strongly depend on ourselves, see [52]. Below I will discuss some factors that may influence the development.

5.1   The Base

Automatic control is a well defined academic discipline. There is a vast amount of intellectually stimulating material that is useful for a wide range of engineering problems. Experiments with automatic control is a very good way to enhance the student's engineering ability. It provides a complex system that includes sensors, actuators, electronics, computers, algorithms.

Our discipline is very young scientifically. Courses have appeared in a partially random fashion and there is currently a lot of diversity. We need to take a careful look at our knowledge base and explore how it can be weeded and streamlined to give a better education both to specialists and to students who only take one or two courses in automatic control.

The importance of trademarks is obvious to anyone who sees how carefully the major companies watch them. In automatic control we use too many names for our discipline, control engineering, feedback control, systems science, and cybernetics. I think it is important to pick one word for our discipline and stick to it. We should probably also consider the academic positioning of the field more carefully.

5.2   Intellectual Drivers

The health of any discipline depends strongly on if we can attract the brightest students to it. This is easy to find out at each university. Stimulating and challenging problems is the thing that attracts the bright students. We must ensure that we are indeed working on those problems; there are lots of them in the field.

5.3   Relations to Other Fields

Automatic control has a strong base in mathematics. The interaction goes in two directions, a wide range of mathematics has found its use in control, but control has also stimulated the development of mathematics. Typical examples are optimal control, which revitalized the field of variational calculus, stochastic control, and the recent work on nonlinear control systems. A more detailed discussion on this is found in [20].

There are also many relations to computer science. Computing is used extensively for analysis and design, computers are key elements in implementation of control systems. There are also many areas where there is cross fertilization in methodology, see [9].

There are also many interactions between automatic control and specialized disciplines that relate to the applications of control. It is, of course, very important to have a good understanding of the systems to be controlled. This is particularly important when the integration of control design and process design is considered.

5.4 Balance of Theory and Applications

The balance between theory and practice is another key issue. Throughout the short history of our field there has been a continuous and sometimes heated debate about the GAP between theory and practice. I personally believe that it is important to strike a balance. It is important to look inwards to improve our basic understanding, our theory and our design methods. But this is not enough; we must also look at the applications. I feel very strongly that automatic control is a systems discipline, which means that we should take full systems responsibility. This means that we should educate students who are capable of solving the whole problem from conceptual design to implementation and commissioning. Control will fade away if we consider the job done when a mathematical description of the control law is obtained. This means that we need to consider implementation, commissioning, and operation.

Lately there have been several instances where control engineers have been very slow in getting into an application area, although we had been in a position to make major contributions. I am a firm believer in theory both as a base for analysis and design and as a good tool for sharpening intellect. The following quote from John van Neumann gives a sobering thought: "As a mathematical discipline travels far from its empirical source, or still more, if it is a second or third generation only indirectly inspired by ideas coming from reality, it is beset with very grave dangers. It becomes more and more aestheticizing, more and more purely l'art pour l'art ... . At a great distance from its source or after much 'abstract' inbreeding, a mathematical subject is in great danger of degeneration." If this applies to mathematics how about our own field?

5.5   Complex Systems

There is a general tendency that the engineering systems are becoming
more and more complex. There are many mechanisms that introduces
complexity; size and interaction are two factors. It is not uncommon
to have chemical processes with many thousand feedback loops. Astro-
nomical telescopes with adaptive optics may have a very large number of
reflecting surfaces whose orientation are controlled individually by feedback systems. In the process industries it is common to use recirculation
to save energy and to reduce pollution. Recirculation introduces a coupling from the output stream to the input stream, which can generate
very complex behavior. Another factor that introduces complexity is
that continuous control systems are often mixed with logic and sequential machines. This is the case even for simple feedback systems with
several operating modes. One example is the system for cruise control
in cars. Other examples are found in process control, where many continuous controllers are combined with systems for logic and sequencing.
Such systems are very difficult to analyze and design. The emerging
field of hybrid systems is one attempt to address these challenges. This
is an interesting meeting place of control and computer science, see [9].

5.6   Integrated Process and Control Design

In automatic control we often take the attitude that the system to be
controlled is given a priori. Such an approach has strong limitations
because there may be inherent limitations to what can be achieved.
Superior systems can often be obtained by designing the complete system
including the process and its control system concurrently. An early
example from 1901 was quoted when discussing flight control in Section
2. Although after the Wright brothers pioneering flights one learned
to build stable and maneuverable aircraft the advantages of inherently
unstable aircrafts has recently received much interest. Such systems
depend critically on a flight control system for stability.

Other examples are found in development of instrumentation where
precision and bandwidth could be improved by several orders of mag-
nitude by a proper combination of feedback control and instrument design. In robotics, substantial gains in performance, precision, and price
are obtained by a light flexible mechanical structure that obtains its
stiffness by an automatic control system. Other examples are found in
mechatronics and power systems.

Very good arguments for an integrated design of a process and its control system are also put forward in [62]. There it is emphasized that much too often the control engineer is faced with a process that is very difficult to control properly. They also advocate that a study of the wide sense controllability of a process should always be done at the design stage. It is very costly to make changes afterwards.

When the automatic control system becomes a critical part of the process it may also become mission critical which means that the system will fail if the control system fails. This induces strong demands on the reliability of the control system. Interesting discussion of the consequences of this are found in [52].

6   Conclusions

Automatic control, which emerged as a discipline about 50 years ago, has had a very dynamic development. It is one of the first systems disciplines that transcends the boundaries of the traditional engineering fields. A wide range of control systems show up in all aspects of modern life. The ideas and methodologies of control have had a strong impact on many other fields. There is also a large number of challenging problems that have the potential of attracting the very best students. It is important that we accept the challenges and in particular that we are willing to deal with the full aspects including theory and applications.

7   References

[1] Karl Johan Åström. *Introduction to Stochastic Control Theory*. Academic Press, New York, 1970. Translated into Russian, Japanese and Chinese.

[2] Karl Johan Åström and Torsten Bohlin. Numerical identification of linear dynamic systems from normal operating records. In *Proc. IFAC Conference on Self-Adaptive Control Systems*, Teddington, UK, 1965.

[3] Karl Johan Åström and Tore Hägglund. *PID Controllers: Theory, Design, and Tuning*. Instrument Society of America, Research Triangle Park, NC, second edition, 1995.

[4] Karl Johan Åström and Björn Wittenmark. *Computer Controlled Systems—Theory and Design*. Prentice-Hall, Englewood Cliffs, New Jersey, second edition, 1990.

[5] Karl Johan Åström and Björn Wittenmark. *Adaptive Control*. Addison-Wesley, Reading, Massachusetts, second edition, 1995.

[6] M. Athans and P. L. Falb. *Optimal Control*. McGraw-Hill, New York, 1966.

[7] R. Bellman. *Dynamic Programming*. Princeton University Press, New Jersey, 1957.

[8] R. Bellman, I. Glicksberg, and O. A. Gross. Some aspects of the mathematical theory of control processes. Technical Report R-313, The RAND Corporation, Santa Monica, Calif., 1958.

[9] Albert Benveniste and Karl Johan Åström. Meeting the challenge of computer science in the industrial applications of control: An introductory discussion to the Special Issue. *Automatica*, 29:1169–1175, 1993.

[10] H. S. Black. Stabilized feedback amplifiers. *Bell System Technical Journal*, 13:1–18, 1934.

[11] H. S. Black. Inventing the negative feedback amplifier. *IEEE spectrum*, pages 55–60, December 1977.

[12] H. W. Bode. Relations between attenuation and phase in feedback amplifier design. *Bell System Technical Journal*, 19:421–454, July 1940.

[13] G. S. Brown and D. P. Campbell. *Principles of Servomechanisms*. Wiley & Sons, New York, 1948.

[14] A. E. Bryson and Y. C. Ho. *Applied Optimal Control Optimization, Estimation and Control*. Blaisdell Publishing Company, 1969.

[15] V. Bush. The differential analyzer. *JFI*, 212(4):447–488, 1931.

[16] H. Chestnut, editor. *Impact of Automatic Control—Present and Future*. VDI/VDE-Gesellschaft, Duesseldorf, 1982.

[17] Harold Chestnut and Robert W. Mayer. *Servomechanisms and Regulating System Design*. Wiley, New York, 1959.

[18] G. Evangelisti. *La regolazione delle turbine idrauliche.* Zanichelli, Bologna, 1947.

[19] A. A. Feldbaum. *Optimal Control Theory.* Academic Press, New York, 1965.

[20] W. H. Fleming, editor. *Future Directions in Control Theory—A Mathematical Perspective.* Society for Industrial and Applied Mathematics, Philadelphia, 1988.

[21] H. L. Hazen. Design and test of a high performance servomechanism. *Ibid*, pages 543–580, 1934.

[22] H. L. Hazen. Theory of servomechanisms. *JFI*, 218:283–331, 1934.

[23] A. Hurwitz. On the conditions under which an equation has only roots with negative real parts. *Mathematische Annalen*, 46:273–284, 1895.

[24] A. Ivanoff. Theoretical foundations of the automatic regulation of temperature. *J. Inst. Fuel*, 7:117–130, 1934. discussion pp. 130–138.

[25] H. M. James, N. B. Nichols, and R. S. Phillips. *Theory of Servomechanisms.* Mc-Graw-Hill, New York, 1947.

[26] Rolf Johansson. *System Modeling and Identification.* Prentice Hall, Englewood Cliffs, New Jersey, 1993.

[27] E. I. Jury. *Sampled-Data Control Systems.* John Wiley, New York, 1958.

[28] R. E. Kalman. Contributions to the theory of optimal control. *Boletin de la Sociedad Matématica Mexicana*, 5:102–119, 1960.

[29] R. E. Kalman. New methods and results in linear prediction and filtering theory. Technical Report 61-1, RIAS, February 1961. 135 pp.

[30] R. E. Kalman. On the general theory of control systems. In *Proceedings first IFAC Congress on Automatic Control, Moscow, 1960*, volume 1, pages 481–492, London, 1961. Butterworths.

[31] R. E. Kalman and R. S. Bucy. New results in linear filtering and prediction theory. *Trans ASME (J. Basic Engineering)*, 83 D:95–108, 1961.

[32] R. E. Kalman, P. L. Falb, and M. A. Arbib. *Topics in Mathematical System Theory*. McGraw-Hill, New York, 1969.

[33] R. E. Kalman, Y. Ho, and K. S. Narendra. *Controllability of Linear Dynamical Systems*, volume 1 of *Contributions to Differential Equations*. John Wiley & Sons, Inc., New York, 1963.

[34] A. N. Kalmogorov. *Interpolation and Extrapolation of Stationary Random Sequences*. Math. 5. Bull. Moscow Univ., USA, 1941.

[35] A. H. Levis. Challenges to control—a collective view report of the workshop held at University of Santa Clara on september 18-19, 1986. *IEEE Trans. Automat. Contr.*, AC-32:274–285, 1987.

[36] L. Ljung. *System Identification—Theory for the User*. Prentice Hall, Englewood Cliffs, New Jersey, 1987.

[37] U. Luoto. *20 Years Old; 20 Years Young. An Anniversary Publication 1957–1977*. Pergamon Press, 1978. IFAC International Federation of Automatic Control.

[38] A. M. Lyapunov. *The General Problem of the Stability of Motion (in Russian)*. Kharkov Mathematical Society (250 pp.), 1892. Collected Works II, 7. Republished by the University of Toulouse 1908 and Princeton University Press 1949 (in French), republished in English by Int. J. Control 1992.

[39] J. C. Maxwell. On governors. *Proceedings of the Royal Society of London*, 16:270–283, 1868. Also published in "Mathematical Trends in Control Theory" edited by R. Bellman and R. Kalaba, Dover Publications, New York 1964, pp. 3–17.

[40] N. Minorsky. Directional stability of automatically steered bodies. *J. Amer. Soc. of Naval Engineers*, 34(2):280–309, 1922.

[41] George C. Newton, Jr, Leonard A. Gould, and James F. Kaiser. *Analytical Design of Linear Feedback Controls*. John Wiley & Sons, 1957.

[42] Harry Nyquist. Regeneration theory. *Bell System Technical Journal*, 11:126–147, 1932. Also published in "Mathematical Trends in Control Theory" edited by R. Bellman and R. Kalaba, Dover Publications, New York 1964, pp. 83–105.

[43] R. C. Oldenbourg and H. Sartorius. *Dynamics of Automatic Controls*. ASME, New York, N. Y., 1948.

[44] R. Oldenburger. *Frequency Response*. The MacMillan Company, New York, 1955.

[45] W. Oppelt. Die Flugzeugkurssteuerung im Geradeausflug. *Luftfahrtforschung*, 14:270–282, 1937.

[46] W. Oppelt. *Kleines Handbuch technischer Regelvorgænge*. Verlag Chemie, 1947.

[47] L. S. Pontryagin, V. G. Boltyanskii, R. V. Gamkrelidze, and E. F. Mischenko. *The Mathematical Theory of Optimal Processes*. John Wiley, New York, 1962.

[48] J. R. Ragazzini and G. F. Franklin. *Sampled-Data Control Systems*. McGraw-Hill, New York, 1958.

[49] E. J. Routh. *A Treatise on the Stability of a Given State of Motion*. Macmillan, London, 1877. Reprinted in Fuller (1975).

[50] Torsten Söderström and Petre Stoica. *System Identification*. Prentice-Hall, London, UK, 1989.

[51] V. V. Solodovnikov. The frequency-response method in the theory of regulation. *Automatica i Telemekhanika*, 8:65–88, 1947.

[52] G. Stein. Respect the unstable. In *The 1989 Bode Lecture*, IEEE, Piscataway, NJ, 1989. IEEE Control Systems Society.

[53] M. Tolle. *Die Regelung der Kraftmaschinen*. Springer-Verlag, Berlin, 1905. 1909, 2nd edition; 1922, 3rd edition.

[54] J. Truxal. *Automatic Feedback Control System Synthesis*. McGraw-Hill, New York, 1955.

[55] H. F. Tsien. *Engineering Cybernetics*. McGraw-Hill Book Company, Inc, New York, 1954.

[56] Ya. Z. Tsypkin. *Theorie der Relais Systeme der Automatischen Regelung*. R. Oldenburg, Munich, Germany, 1958.

[57] A. Tustin, editor. *Automatic and Manual Control: Proceedings of the 1951 Cranfield Conference*, New York, N. Y., 1952. Academic Press.

[58] J. Vyshnegradskii. Sur la théorie générale des régulateurs. *Compt. Rend. Acid. Sci. Paris*, 83:318–321, 1876.

[59] N. Wiener. *The Extrapolation, Interpolation, and Smoothing of Stationary Time Series with Engineering Applications*. Wiley, New York, 1949. Originally issued as a classified MIT Rad. Lab. Report in February, 1942.

[60] L. A. Zadeh and C. A. Desoer. *Series in Systems Science*. McGraw-Hill, Inc., USA, 1963.

[61] J. G. Ziegler and N. B. Nichols. Optimum settings for automatic controllers. *Trans. ASME*, 64:759–768, 1942.

[62] J. G. Ziegler, N. B. Nichols, and N. Y. Rochester. Process lags in automatic-control circuits. *Trans. ASME*, 65:433–444, July 1943.

2. Semiglobal Robust Regulation of Nonlinear Systems

Alberto Isidori*

Abstract

This paper describes a recent approach to the problem of output regulation of nonlinear systems, which unifies and extends a number of earlier results on the subject. In particular the approach in question yields necessary and sufficient conditions for local structurally stable regulation and, for special classes of nonlinear systems, also semiglobal robust regulation.

1 Introduction

The problem of output regulation (or, what is the same, the servomechanism problem) is to control the output of a system so as to achieve asymptotic tracking of prescribed trajectories and/or asymptotic rejection of unwanted disturbances. A typical setup in which the problem is usually posed is the one in which the exogenous inputs, namely commands and disturbances, may range over the set of all possible trajectories of a given autonomous linear, time-invariant and finite dimensional system, commonly known as the *exosystem*. For linear, time-invariant and finite dimensional plants the problem was completely solved in the works of Davison, Francis and Wonham [3], [7], [8]. In the case of *nonlinear*, time-invariant and finite dimensional plants, the problem was studied in [8], [9], [10], [15], [12], [13].

*Dipartimento di Informatica e Sistemistica, Università di Roma "La Sapienza", 00184 Rome, and Department of Systems Science and Mathematics, Washington University, St.Louis, MO 63130.

All these contributions, however, only consider (with the exception of those in [8] and [10], dealing with constant exogenous inputs) the ideal case of a controlled plant not affected by uncertainties. More recently, several authors have addressed the problems of achieving either *structurally stable* regulation (i.e. regulation properties preserved under small parameter variations) or *robust* regulation (i.e. regulation properties preserved under parameter variations ranging on *a priori fixed* intervals). The first major contribution in the direction of solving the problem of dealing with parameter uncertainties was the idea by Khalil [17] (see also [18]) of synthesizing an internal model able to generate not just the trajectories of the exosystem, but also a number of their "higher order" nonlinear "deformations". In particular, the paper [18] uses this approach to the design of an internal model, and the approach of [5] to semiglobal stabilization via output feedback, in order to solve the problem of semiglobal output regulation for a class of nonlinear systems. Related results can also be found in [21]. The search for an appropriate internal model able to cope with perturbations in a nonlinear system was also independently pursued by Huang and Lin, in [11] and by Delli Priscoli in [4]. However these papers deal only with "small" perturbations (i.e. structurally stable regulation) and local convergence, as opposite to [18] which deals with perturbations ranging in a prescribed set and semiglobal convergence.

A somewhat more general approach to the problem of structurally stable regulation was presented in Chapter 8 of [14]. This approach unifies all earlier results on this subject and also provides a precise set of necessary and sufficient conditions, which appeal to the notion of *immersion* of a system into another system. The purpose of this paper is to illustrate the applicability of this approach to a significant class of nonlinear systems.

The systems considered throughout this chapter are modeled by equations of the form

$$\begin{aligned}
\dot{x} &= f(x, u, w, \mu) \\
e &= h(x, w, \mu)
\end{aligned} \tag{1}$$

with state $x \in \mathbb{R}^n$, control input $u \in \mathbb{R}^m$, regulated output $e \in \mathbb{R}^m$, subject to an exogenous disturbance input $w \in \mathbb{R}^d$, in which $\mu \in \mathcal{P} \subset \mathbb{R}^p$ is a vector of unknown parameters and $\mathcal{P}$ is compact set. $f(x, u, w, \mu)$ and $h(x, w, \mu)$ are C^k functions of their arguments (for some large k), and $f(0, 0, 0, \mu) = 0$ and $h(0, 0, \mu) = 0$ for each value of μ. Without loss of generality, we suppose $0 \in \text{int}(\mathcal{P})$. The exosystem

$$\dot{w} = Sw \tag{2}$$

is assumed to be *neutrally stable*.

The control input to (1) is to be provided by a controller modeled by equations of the form

$$\begin{aligned}
\dot{\xi} &= \eta(\xi, e) \\
u &= \theta(\xi)
\end{aligned} \tag{3}$$

with state $\xi \in \mathbb{R}^\nu$, in which $\eta(\xi, e)$ and $\theta(\xi)$ are C^k functions of their arguments, and $\eta(0,0) = 0$, $\theta(0) = 0$.

2 Structurally stable regulation

In this section we describe how to solve the so-called problem of *structurally stable* nonlinear output regulation. The problem in question is defined as follows.

Definition The controller (3) is said to solve the problem of *structurally stable* output regulation for the plant (1) with exosystem (2) if there exists a neighborhood $\mathcal{P}^\circ$ of $\mu = 0$ in $\mathbb{R}^p$ such that, for each $\mu \in \mathcal{P}^\circ$:

(a) the unforced closed loop system

$$\begin{aligned}
\dot{x} &= f(x, \theta(\xi), 0, \mu) \\
\dot{\xi} &= \eta(\xi, h(x, 0, \mu))
\end{aligned}$$

has a locally exponentially stable equilibrium at $(x, \xi) = (0, 0)$,

(b) the forced closed loop system

$$\begin{aligned}
\dot{x} &= f(x, \theta(\xi), w, \mu) \\
\dot{\xi} &= \eta(\xi, h(x, w, \mu)) \\
\dot{w} &= Sw
\end{aligned}$$

is such that

$$\lim_{t \to \infty} e(t) = 0$$

for each initial condition $(x(0), \xi(0), w(0))$ in some neighborhood of the equilibrium $(0, 0, 0)$. ◁

In order to present the necessary and sufficient conditions for the existence of a solution of the problem of structurally stable output regulation, it is convenient to recall first the notion of *immersion* of a system into another system, introduced by Fliess in [6]. To the purpose of the present analysis, the notion in question reduces to the following one: the autonomous system with output

$$\dot{x} = f(x), \qquad y = h(x) \tag{4}$$

is immersed into the autonomous system with output

$$\dot{\xi} = \phi(\xi), \qquad y = \gamma(\xi) \tag{5}$$

if there exists a C^1 mapping $\xi = \tau(x)$ such that, for every x^0, the output produced by (4) in the initial state $x(0) = x^0$ and the output produced by (5) in the initial state $\xi(0) = \tau(x^0)$ are *identical*.

The following result provides necessary and sufficient conditions for the existence of solutions of the problem of structurally stable output regulation. Set

$$A(\mu) = \left[\frac{\partial f}{\partial x}\right]_{(0,0,0,\mu)}, \qquad B(\mu) = \left[\frac{\partial f}{\partial u}\right]_{(0,0,0,\mu)}, \qquad C(\mu) = \left[\frac{\partial h}{\partial x}\right]_{(0,0,\mu)}.$$

Theorem 1 *Suppose the exosystem (2) is neutrally stable. There exists a solution of the problem of structurally stable output regulation if and only if there exist mappings $x = \pi^{\mathrm{a}}(w, \mu)$ and $u = c^{\mathrm{a}}(w, \mu)$, with $\pi^{\mathrm{a}}(0, \mu) = 0$ and $c^{\mathrm{a}}(0, \mu) = 0$, both defined in a neighborhood $W^\circ \times P^\circ \subset \mathbb{R}^d \times P$ of the origin, satisfying the conditions*

$$\frac{\partial \pi^{\mathrm{a}}(w, \mu)}{\partial w} Sw = f(\pi^{\mathrm{a}}(w, \mu), c^{\mathrm{a}}(w, \mu), w, \mu) \tag{6}$$
$$0 = h(\pi^{\mathrm{a}}(w, \mu), w, \mu)$$

for all $(w, \mu) \in W^\circ \times P^\circ$, and such that the autonomous system with output

$$\dot{w} = Sw$$
$$\dot{\mu} = 0 \tag{7}$$
$$u = c^{\mathrm{a}}(w, \mu)$$

is immersed into a system

$$\dot{\xi} = \varphi(\xi)$$
$$u = \gamma(\xi),$$

defined on a neighborhood Ξ° of the origin in $\mathbb{R}^\nu$, in which $\varphi(0) = 0$ and $\gamma(0) = 0$ and the two matrices

$$\Phi = \left[\frac{\partial \varphi}{\partial \xi}\right]_{\xi=0}, \qquad \Gamma = \left[\frac{\partial \gamma}{\partial \xi}\right]_{\xi=0}$$

are such that the pair

$$\begin{pmatrix} A(0) & 0 \\ NC(0) & \Phi \end{pmatrix}, \qquad \begin{pmatrix} B(0) \\ 0 \end{pmatrix} \tag{8}$$

is stabilizable for some choice of the matrix N, and the pair

$$(C(0) \quad 0), \qquad \begin{pmatrix} A(0) & B(0)\Gamma \\ 0 & \Phi \end{pmatrix} \tag{9}$$

is detectable.

Remark. Note that the linear approximation of (7) at the equilibrium $(w, \mu) = (0, 0)$ cannot be detectable. In fact, since $c^{\mathrm{a}}(0, \mu) = 0$ by hypothesis,

$$\frac{\partial}{\partial \mu} c^{\mathrm{a}}(0, \mu) = 0,$$

and the linear approximation in question is characterized by a pair of matrices of the form

$$(* \quad 0), \qquad \begin{pmatrix} S & 0 \\ 0 & 0 \end{pmatrix}$$

which is indeed not detectable. Thus, it is not possible to have the conditions of the Theorem directly satisfied by the trivial immersion of (7) into itself. However, as shown below, (7) may be immersed into *another* system, having a detectable approximation at $\xi = 0$.◁

Remark. The condition that system (7) is immersed into a system

$$\begin{aligned} \dot{\xi} &= \varphi(\xi) \\ u &= \gamma(\xi), \end{aligned}$$

is the existence of a mapping $\tau^{\mathrm{a}}(w, \mu)$ such that

$$\frac{\partial \tau^{\mathrm{a}}}{\partial w} Sw = \varphi(\tau^{\mathrm{a}}(w, \mu)), \qquad c^{\mathrm{a}}(w, \mu) = \gamma(\tau^{\mathrm{a}}(w, \mu)).$$

Choose N so that the pair (8) is stabilizable and K, L, M so that

$$\left(\begin{pmatrix} A(0) & B(0)\Gamma \\ NC(0) & \Phi \\ L\,(C(0) & 0) \end{pmatrix} \quad \begin{pmatrix} B(0) \\ 0 \\ K \end{pmatrix} M \right)$$

has all eigenvalues with negative real part. Then, the controller

$$\begin{aligned}
\dot{\xi}_0 &= K\xi_0 + Le \\
\dot{\xi}_1 &= \varphi(\xi_1) + Ne \\
u &= M\xi_0 + \gamma(\xi_1) \; .
\end{aligned} \tag{10}$$

solves the problem of structurally stable output regulation. In fact, it is immediate to see that the unforced closed loop system

$$\begin{aligned}
\dot{x} &= f(x, M\xi_0 + \gamma(\xi_1), 0, \mu) \\
\dot{\xi}_0 &= K\xi_0 + Lh(x, 0, \mu) \\
\dot{\xi}_1 &= \varphi(\xi_1) + Nh(x, 0, \mu)
\end{aligned}$$

has a locally exponentially stable equilibrium at $(x, \xi_0, \xi_1) = (0, 0, 0)$ for all μ in some neighborhood of $\mu = 0$. Moreover, by construction, the set

$$M_c = \{(x, \xi_0, \xi_1, w, \mu) : x = \pi^a(w, \mu), \xi_0 = 0, \xi_1 = \tau^a(w, \mu)\}$$

is a center manifold for the system

$$\begin{aligned}
\dot{x} &= f(x, M\xi_0 + \gamma(\xi_1), w, \mu) \\
\dot{\xi}_0 &= K\xi_0 + Lh(x, w, \mu) \\
\dot{\xi}_1 &= \varphi(\xi_1) + Nh(x, w, \mu) \\
\dot{w} &= Sw \\
\dot{\mu} &= 0.
\end{aligned}$$

at the equilibrium $(x, \xi_0, \xi_1, w, \mu) = (0, 0, 0, 0, 0)$. Thus, $\lim_{t \to \infty} e(t) = 0$ for each initial condition $(x(0), \xi_0(0), \xi_1(0), w(0))$ in some neighborhood of $(0, 0, 0, 0)$ for all μ in some neighborhood of $\mu = 0$. ◁

This result contains a number of relevant corollaries. Let $L_s\lambda(w, \mu)$ denote the derivative of a function $\lambda(w, \mu)$ along Sw

$$L_s\lambda(w, \mu) = \frac{\partial \lambda(w, \mu)}{\partial w} Sw \; .$$

Corollary 1 *Suppose there exist mappings $x = \pi^a(w, \mu)$ and $u = c^a(w, \mu)$, with $\pi^a(0, \mu) = 0$ and $c^a(0, \mu) = 0$, both defined in a neighborhood $W^\circ \times \mathcal{P}^\circ \subset \mathbb{R}^d \times \mathcal{P}$ of the origin, satisfying the conditions (6). Suppose also there exists integers $p_1, \ldots, p_m$ and functions*

$$
\begin{aligned}
\phi_i \quad &: \mathbb{R}^{p_i} \quad \rightarrow \quad \mathbb{R} \\
(\zeta_1, \ldots, \zeta_i) \quad &\mapsto \quad \phi_i(\zeta_1, \ldots, \zeta_i)
\end{aligned}
$$

such that, for all $1 \leq i \leq m$, the i-th component $c_i^a(w, \mu)$ of $c^a(w, \mu)$ satisfies

$$
L_s^{p_i} c_i^a(w, \mu) = \phi_i(c_i^a(w, \mu), L_s c_i^a(w, \mu), \ldots, L_s^{p_i-1} c_i^a(w, \mu)), \tag{11}
$$

for all $(w, \mu) \in W^\circ \times \mathcal{P}^\circ$. Set

$$
d_{ij} = \left[\frac{\partial \phi_i}{\partial \zeta_j} \right]_{(0, \ldots, 0)}
$$

and

$$
d_i(\lambda) = d_{i0} + d_{i1}\lambda + \ldots + d_{ip_i-1}\lambda^{p_i-1} - \lambda^{p_i} .
$$

Suppose the pair $(A(0), B(0))$ is stabilizable, the pair $(C(0), A(0))$ is detectable, and the matrix

$$
\begin{pmatrix} A(0) - \lambda I & B(0) \\ C(0) & 0 \end{pmatrix} \tag{12}
$$

is nonsingular for every λ which is a root of any of the polynomials $d_1(\lambda), \ldots, d_m(\lambda)$ having non-negative real part. Then there exists a controller which solves the problem of structurally stable local output regulation.

Corollary 2 *Suppose there exist mappings $x = \pi^a(w, \mu)$ and $u = c^a(w, \mu)$, with $\pi^a(0, \mu) = 0$ and $c^a(0, \mu) = 0$, both defined in a neighborhood $W^\circ \times \mathcal{P}^\circ \subset \mathbb{R}^d \times \mathcal{P}$ of the origin, satisfying the conditions (6). Suppose also that, some set of q real numbers $a_0, a_1, \ldots, a_{q-1}$,*

$$
L_s^q c^a(w, \mu) = a_0 c^a(w, \mu) + a_1 L_s c^a(w, \mu) + \cdots + a_{q-1} L_s^{q-1} c^a(w, \mu), \tag{13}
$$

for all $(w, \mu) \in W^\circ \times \mathcal{P}^\circ$. Suppose the pair $(A(0), B(0))$ is stabilizable, the pair $(C(0), A(0))$ is detectable, and the matrix (12) is nonsingular for every λ which is a root of the polynomial

$$
p(\lambda) = a_0 + a_1\lambda + \ldots + a_{q-1}\lambda^{q-1} - \lambda^q \tag{14}
$$

with non-negative real part. Then there exists a linear controller which solves the problem of structurally stable local output regulation.

It is interesting to observe that Corollary 2 contains, as particular cases, the results about structurally stable regulation of linear and non-linear systems of [7], [3], [10], [11] and [4]. Note also that (13) is a *necessary* condition for the existence of a linear controller which solves the problem of structurally stable output regulation.

3   The special case of systems in triangular form

We describe in this section some special classes of nonlinear single-input single-output system for which the condition for structurally stable local regulation illustrated in the previous sections can easily be tested. Consider, for example, systems which can be described by equations of the following form

$$
\begin{aligned}
\dot{x}_1 &= a_2(\mu)x_2 + p_1(x_1, w, \mu) \\
\dot{x}_2 &= a_3(\mu)x_3 + p_2(x_1, x_2, w, \mu) \\
&\cdots \\
\dot{x}_{n-1} &= a_n(\mu)x_n + p_{n-1}(x_1, x_2, \ldots, x_{n-1}, w, \mu) \\
\dot{x}_n &= p_n(x_1, x_2, \ldots, x_n, w, \mu) + b(\mu)u \\
e &= c(\mu)x_1 + q(w, \mu) \,,
\end{aligned}
\tag{15}
$$

in which, for each $\mu \in \mathcal{P}$, the coefficients $a_2(\mu)$, $a_3(\mu)$, ..., $a_n(\mu)$, $b(\mu)$, $c(\mu)$ are nonzero, $q(0, \mu) = 0$ and $p_i(0, \ldots, 0, 0, \mu) = 0$ for $1 \le i \le n$. The system in question is a system of the form

$$
\begin{aligned}
\dot{x} &= F(\mu)x + G(\mu)u + P(x, w, \mu) \\
e &= H(\mu)x + q(x, \mu)
\end{aligned}
\tag{16}
$$

with

$$
F(\mu) = \begin{pmatrix}
0 & a_2(\mu) & 0 & \cdots & 0 \\
0 & 0 & a_3(\mu) & \cdots & 0 \\
\cdot & \cdot & \cdot & \cdots & \cdot \\
0 & 0 & 0 & \cdots & a_n(\mu) \\
0 & 0 & 0 & \cdots & 0
\end{pmatrix}, \quad B(\mu) = \begin{pmatrix}
0 \\
0 \\
\cdot \\
0 \\
b(\mu)
\end{pmatrix}
$$

$$
H(\mu) = \begin{pmatrix} c(\mu) & 0 & 0 & \cdots & 0 \end{pmatrix} \,,
$$

and the vector $P(x, w, \mu)$ exhibits a triangular dependence on the individual components of x. In particular, its relative degree is equal to n, for all μ.

For this system it is immediate to find a solution of the regulator equations (6), which is defined for all w, μ in $\mathbb{R}^d \times \mathcal{P}$. In fact, observe that the second equation of (6), i.e.

$$h(\pi^{\mathrm{a}}(w, \mu), w, \mu) = 0 \, ,$$

directly provides the expression of first component $\pi_1^{\mathrm{a}}(w, \mu)$ of $\pi^{\mathrm{a}}(w, \mu)$, which is

$$\pi_1^{\mathrm{a}}(w, \mu) = -\frac{q(w, \mu)}{c(\mu)} \, .$$

Because of the special structure of the system (15), the first equation of (6) uniquely determines all other components of $\pi^{\mathrm{a}}(w, \mu)$. In particular, its second component $\pi_2^{\mathrm{a}}(w, \mu)$ is determined by the identity

$$\frac{\partial \pi_1^{\mathrm{a}}}{\partial w} S w = a_2(\mu) \pi_2^{\mathrm{a}}(w, \mu) + p_1(\pi_1^{\mathrm{a}}(w, \mu), w, \mu) \, ,$$

which yields

$$\pi_2^{\mathrm{a}}(w, \mu) = \frac{1}{a_2(\mu)} \left(\frac{\partial \pi_1^{\mathrm{a}}}{\partial w} S w - p_1(\pi_1^{\mathrm{a}}(w, \mu), w, \mu) \right) \, .$$

This procedure can be iterated up to the last equation, which eventually provides the unique expression for $c^{\mathrm{a}}(w, \mu)$, which is

$$c^{\mathrm{a}}(w, \mu) = \frac{1}{b(\mu)} \left(\frac{\partial \pi_n^{\mathrm{a}}}{\partial w} S w - p_n(\pi_1^{\mathrm{a}}(w, \mu), \ldots, \pi_n^{\mathrm{a}}(w, \mu), w, \mu) \right) \, .$$

If the expression of $c^{\mathrm{a}}(w, \mu)$ is such that, for some integer p and some function $\phi(\xi_1, \xi_2, \ldots, \xi_p)$,

$$L_s^p c^{\mathrm{a}}(w, \mu) = \phi(c^{\mathrm{a}}(w, \mu), L_s c^{\mathrm{a}}(w, \mu), \ldots, L_s^{p-1} c^{\mathrm{a}}(w, \mu)) \, , \qquad (17)$$

then the autonomous system with output (7) is immersed into a nonlinear system

$$\dot{\xi} = \varphi(\xi)$$
$$u = \gamma(\xi) \, ,$$

in which

$$\varphi(\xi) = \begin{pmatrix} \xi_2 \\ \xi_3 \\ \cdots \\ \xi_p \\ \phi(\xi_1, \xi_2, \cdots, \xi_p) \end{pmatrix} \, , \qquad \gamma(\xi) = \xi_1 \, .$$

Observe that the matrices $A(\mu), B(\mu), C(\mu)$ which characterize the linear approximation of (16) at the equilibrium $x = 0$ have the following structure

$$A(\mu) = F(\mu) + \left[\frac{\partial P(x, \mu, w)}{\partial x}\right]_{0,0,0}, \quad B(\mu) = G(\mu), \quad C(\mu) = H(\mu),$$

where

$$\left[\frac{\partial P(x, \mu, w)}{\partial x}\right]_{0,0,0}$$

is a lower triangular matrix. Thus, in view of the special structure of $F(\mu), G(\mu), H(\mu)$, for all $\mu \in \mathcal{P}$ the pair $(A(\mu), B(\mu))$ is controllable, the pair $(C(\mu), A(\mu))$ is stabilizable, the matrix

$$\begin{pmatrix} A(\mu) - \lambda I & B(\mu) \\ C(\mu) & 0 \end{pmatrix}$$

is nonsingular for all λ. Thus all the conditions for the existence of a controller solving the problem of structurally stable local output regulation are fulfilled.

These considerations can be extended to more general classes of systems, for instance to systems described by equations of the form

$$
\begin{aligned}
\dot{z} &= f_0(z, x_1, w, \mu) \\
\dot{x}_1 &= a_2(\mu)x_2 + p_1(z, x_1, w, \mu) \\
\dot{x}_2 &= a_3(\mu)x_3 + p_2(z, x_1, x_2, w, \mu) \\
&\quad \cdots \\
\dot{x}_{r-1} &= a_r(\mu)x_r + p_{r-1}(z, x_1, x_2, \ldots, x_{r-1}, w, \mu) \\
\dot{x}_r &= b(\mu)u + p_r(z, x_1, x_2, \ldots, x_r, w, \mu) \\
e &= c(\mu)x_1 + q(w, \mu),
\end{aligned} \tag{18}
$$

provided that there exists a unique and globally defined mapping $z = \zeta(w, \mu)$ satisfying

$$\frac{\partial \zeta}{\partial w}Sw = f_0(\zeta(w, \mu), \frac{1}{c(\mu)}q(w, \mu), w, \mu) \tag{19}$$

for all $(w, \mu) \in \mathbb{R}^d \times \mathcal{P}$. If this is the case, in fact, the regulator equations can be recursively solved in essentially the same manner. After having split $\pi^{\mathrm{a}}(w, \mu)$ as

$$\pi^{\mathrm{a}}(w, \mu) = \begin{pmatrix} \zeta(w, \mu) \\ \pi_1(w, \mu) \\ \pi_2(w, \mu) \\ \cdots \\ \pi_r(w, \mu) \end{pmatrix},$$

it is immediate to realize that the second equation of (6) determines uniquely the component $\pi_1(w, \mu)$, that the identity (19) determines uniquely $\zeta(w, \mu)$ and that the latter determines uniquely, via the first of (6), all remaining components $\pi_2(w, \mu), \ldots, \pi_r(w, \mu)$ of $\pi^a(w, \mu)$ and $c^a(w, \mu)$.

Remark. A *special case* in which a global solution of (19) exists and the corresponding solutions $\pi^a(w, \mu)$ and $c^a(w, \mu)$ of (6) satisfy the condition (17) for immersion into a system having a detectable linear approximation, is when

$$f_0(z, x_1, w, \mu) = Z(\mu)z + p_0(x_1, w, \mu)$$

and, for each $\mu \in \mathcal{P}$, $q(w, \mu), p_0(x_1, w, \mu), p_1(z, x_1, w, \mu), \ldots, p_r(z, x_1, x_2, \ldots, x_r, w, \mu)$ are polynomials in $z, x_1, x_2, \ldots, x_r, w$ of degree not exceeding a fixed number k, independent of μ, and a certain restriction is placed on the eigenvalues of the matrix $Z(\mu)$.

Observe, in fact, that in this case $\pi_1(w, \mu)$ is a polynomial in w, whose degree does not exceed a fixed number independent of μ. Substituting this function in (19) yields an equation of the form

$$\frac{\partial \zeta}{\partial w} Sw = Z(\mu)\zeta(w, \mu) + Y(w, \mu) \tag{20}$$

in the unknown $\zeta(w, \mu)$, where $Y(w, \mu)$ is a polynomial in w, whose degree does not exceed a fixed number (independent of μ). Let k denote this number. Let $\mathcal{P}_k$ denote the set of all polynomials $p(w)$ of degree less than or equal to k in the variables $w_1, w_2, \ldots, w_d$ with coefficients in $\mathbb{R}$ and satisfying $p(0) = 0$. $\mathcal{P}_k$ is indeed a finite-dimensional vector space over $\mathbb{R}$. If $p(w) \in \mathcal{P}_k$, then

$$L_s p(w) = \frac{\partial p}{\partial w} Sw$$

is still a polynomial in $\mathcal{P}_k$. Thus, the mapping

$$\begin{aligned} D_k : \mathcal{P}_k &\rightarrow \mathcal{P}_k \\ p(w) &\mapsto \frac{\partial p}{\partial w} Sw \end{aligned} \tag{21}$$

is an $\mathbb{R}$-linear mapping from a finite dimensional vector space to itself. Using this notation, observe that if the entries of $\zeta(w, \mu)$ are in $\mathcal{P}_k$, the left-hand side of (20) can be written in the form

$$\frac{\partial \zeta}{\partial w} Sw = \begin{pmatrix} D_k & 0 & \cdots & 0 \\ 0 & D_k & \cdots & 0 \\ \cdot & \cdot & \cdots & \cdot \\ 0 & 0 & \cdots & D_k \end{pmatrix} \zeta(w, \mu) .$$

Thus, it is easy to conclude that (20) has a solution $\zeta(w, \mu)$, whose components are polynomials of degree not exceeding k, if none of the roots of the minimal polynomial of D_k are eigenvalues of the matrix $Z(\mu)$. In fact, equation (20) can be seen as a Sylvester equation which, if the minimal polynomial of D_k and that of $Z(\mu)$ have no common roots, has a unique solution $\zeta(w, \mu)$ whose components are polynomials of degree not exceeding k.

From this, the recursive solution of the equations (6) presented above yields, at each step, a component of $\pi^{\mathrm{a}}(w, \mu)$ which is a polynomial in w, and whose degree does not exceed a fixed number independent of μ. Thus, eventually, also the unique expression found for $c^{\mathrm{a}}(w, \mu)$ is a polynomial in w, whose degree does not exceed a fixed number. This, being the case, it is immediate to conclude that there exist numbers $a_0, a_1, \ldots, a_{q-1}$ such that

$$L_s^q c^{\mathrm{a}}(w, \mu) = a_0 c^{\mathrm{a}}(w, \mu) + a_1 L_s c^{\mathrm{a}}(w, \mu) + \cdots + a_{q-1} L_s^{q-1} c^{\mathrm{a}}(w, \mu) , \quad (22)$$

for all $(w, \mu) \in \mathbb{R}^d \times \mathcal{P}$, i.e. that the condition (17) is fulfilled.

As far as the remaining conditions of Corollary 2 are concerned, it suffices to assume that all the eigenvalues of $F(\mu)$ have negative real part. In this case, again, for all $\mu \in \mathcal{P}$ the pair $(A(\mu), B(\mu))$ is controllable, the pair $(C(\mu), A(\mu))$ is stabilizable and the matrix

$$\begin{pmatrix} A(\mu) - \lambda I & B(\mu) \\ C(\mu) & 0 \end{pmatrix}$$

is nonsingular for all λ which are roots of the polynomial (14). ◁

4   A globally defined error-zeroing invariant manifold

Motivated by the analysis presented in the previous section, we consider henceforth output regulation problems for systems modeled by equations of the form

$$\begin{aligned}
\dot{z} &= Z(\mu)z + p_0(x_1, w, \mu) \\
\dot{x} &= Fx + Gu + P(z, x, w, \mu) \\
e &= Hx - q(w, \mu)
\end{aligned} \qquad (23)$$

where

$$F = \begin{pmatrix}
0 & 1 & 0 & \cdots & 0 \\
0 & 0 & 1 & \cdots & 0 \\
\cdot & \cdot & \cdot & \cdots & \cdot \\
0 & 0 & 0 & \cdots & 1 \\
0 & 0 & 0 & \cdots & 0
\end{pmatrix}, \quad G = \begin{pmatrix}
0 \\
0 \\
\cdot \\
0 \\
1
\end{pmatrix}$$

$$H = (1 \quad 0 \quad 0 \quad \cdots \quad 0)$$

and

$$P(z, x, w, \mu) = \begin{pmatrix}
p_1(z, x_1, w, \mu) \\
p_2(z, x_1, x_2, w, \mu) \\
\cdots \\
p_{r-1}(z, x_1, x_2, \ldots, x_{r-1}, w, \mu) \\
p_r(z, x_1, x_2, \ldots, x_r, w, \mu)
\end{pmatrix}.$$

In what follows its is assumed that, for each $\mu \in \mathcal{P}$, $q(w, \mu)$, $p_0(x_1, w, \mu)$ $p_1(z, x_1, w, \mu), \ldots, p_r(z, x_1, x_2, \ldots, x_r, w, \mu)$ are polynomials in z, x_1, x_2, $\ldots$, x_r, w of degree not exceeding a fixed number k, independent of μ. Moreover, it is assumed that the eigenvalues of $F(\mu)$ have negative real part, for all $\mu \in \mathcal{P}$.

Under these hypotheses, the equations (6) have a unique and globally defined solution $\pi^a(w, \mu)$, $c^a(w, \mu)$ in which $c^a(w, \mu)$ satisfies an identity of the form (13). This, in turn, uniquely determines the existence of a $q \times q$ matrix Φ, a $1 \times q$ row vector Γ, and a globally defined mapping $\tau(w, \mu)$ such that

$$\begin{aligned}
\frac{\partial \tau^a(w, \mu)}{\partial w} Sw &= \Phi \tau^a(w, \mu) \\
c^a(w, \mu) &= \Gamma \tau^a(w, \mu).
\end{aligned} \qquad (24)$$

As a matter of fact, this occurs for

$$\Phi = \begin{pmatrix}
0 & 1 & 0 & \cdots & 0 \\
0 & 0 & 1 & \cdots & 0 \\
\cdot & \cdot & \cdot & \cdots & \cdot \\
0 & 0 & 0 & \cdots & 1 \\
-a_0 & -a_1 & -a_2 & \cdots & -a_{q-1}
\end{pmatrix}, \quad \tau^a(w, \mu) = \begin{pmatrix}
c^a(w, \mu) \\
L_s c^a(w, \mu) \\
\cdot \\
L_s^{q-2} c^a(w, \mu) \\
L_s^{q-1} c^a(w, \mu)
\end{pmatrix}$$

$$\Gamma = (1 \quad 0 \quad 0 \quad \cdots \quad 0).$$

Consider now a feedback law of the form

$$
\begin{aligned}
\dot{\xi}_0 &= K\xi_0 + Le \\
\dot{\xi}_1 &= \Phi\xi_1 + Ne \\
u &= \alpha(\xi_0) + T\xi_1 .
\end{aligned}
\tag{25}
$$

in which $\alpha(\xi_0)$ is a (possibly nonlinear) function, smooth in a neighborhood of $\xi_0 = 0$ and such that $\alpha(0) = 0$.

Then, it is possible to prove that the following property holds.

Proposition 1 *Suppose (25) asymptotically stabilizes the linear approximation of (23) at the equilibrium point $(\xi_0, \xi_1, z, x) = (0, 0, 0, 0)$, $(w, \mu) = (0, 0)$. Suppose $\sigma(K) \in \mathbf{C}^-$. Then, there exists a $q \times q$ matrix Π satisfying*

$$
\begin{aligned}
\Phi\Pi &= \Pi\Phi \\
T\Pi &= \Gamma ,
\end{aligned}
\tag{26}
$$

where Φ and Γ are defined as in (24). As a consequence, the composite system

$$
\begin{aligned}
\dot{\xi}_0 &= K\xi_0 + L(Hx - q(w, \mu)) \\
\dot{\xi}_1 &= \Phi\xi_1 + N(Hx - q(w, \mu)) \\
\dot{z} &= Z(\mu)z + p_0(x_1, w, \mu) \\
\dot{x} &= Fx + G(\alpha(\xi_0) + T\xi_1) + P(z, x, w, \mu) \\
\dot{w} &= Sw
\end{aligned}
\tag{27}
$$

has a globally defined center manifold

$$
\mathcal{M}_c = \{(\xi_0, \xi_1, z, x, w) : \xi_0 = 0, \xi_1 = \Pi\tau^a(w, \mu), z = \zeta(w, \mu), x = \pi^a(w, \mu)\}
$$

at $(\xi_0, \xi_1, z, x, w) = (0, 0, 0, 0, 0)$.

Proof Note that the linear approximation of (27) at $(\xi_0, \xi_1, z, x, w) = (0, 0, 0, 0, 0, 0)$, is a system of the form

$$
\begin{aligned}
\dot{\xi}_0 &= K\xi_0 + LHx + LQ(\mu)w \\
\dot{\xi}_1 &= \Phi\xi_1 + NHx + NQ(\mu)w \\
\dot{z} &= Z(\mu)z + T_x(\mu)Hx + T_w(\mu)w \\
\dot{x} &= Fx + G(M\xi_0 + T\xi_1) + P_z(\mu)z + P_x(\mu)x + P_w(\mu)w \\
\dot{w} &= Sw .
\end{aligned}
\tag{28}
$$

By hypothesis, the matrix

$$
\begin{pmatrix}
K & 0 & 0 & LH \\
0 & \Phi & 0 & NH \\
0 & 0 & Z(0) & T_x(0)H \\
GM & GT & P_z(0) & F + P_x(0)
\end{pmatrix}
\tag{29}
$$

has all eigenvalues with negative real part. Since the matrix Φ has all eigenvalues on the imaginary axis, the Sylvester equation

$$
\begin{pmatrix} K & 0 & 0 & LH \\ 0 & \Phi & 0 & NH \\ 0 & 0 & Z(0) & T_x(0)H \\ GM & GT & P_z(0) & F+P_x(0) \end{pmatrix} \begin{pmatrix} X_0 \\ X_1 \\ X_z \\ X_x \end{pmatrix} - \begin{pmatrix} X_0 \\ X_1 \\ X_z \\ X_x \end{pmatrix} \Phi = \begin{pmatrix} 0 \\ Y \\ 0 \\ 0 \end{pmatrix}
$$

has a unique solution for any choice of Y. Thus, in particular, for any Y there exist X_1 and X_x such that

$$
\Phi X_1 + NHX_x - X_1\Phi = Y .
$$

From this, using standard properties (see e.g. [20] pages 34–36), it is immediate to conclude that an identity of the form

$$
\Phi X_1 + NHX_x - X_1\Phi = 0
$$

necessarily implies

$$
\Phi X_1 - X_1\Phi = 0 \qquad \text{and} \qquad HX_x = 0 .
$$

Using again the hypotheses that the matrix (29) has all eigenvalues with negative real part and the matrix Φ has all eigenvalues on the imaginary axis, note that the Sylveseter equation

$$
\begin{pmatrix} K & 0 & 0 & LH \\ 0 & \Phi & 0 & NH \\ 0 & 0 & Z(0) & T_x(0)H \\ GM & GT & P_z(0) & F+P_x(0) \end{pmatrix} \begin{pmatrix} X_0 \\ \Pi \\ X_z \\ X_x \end{pmatrix} - \begin{pmatrix} X_0 \\ \Pi \\ X_z \\ X_x \end{pmatrix} \Phi = \begin{pmatrix} 0 \\ 0 \\ 0 \\ G\Gamma \end{pmatrix}
$$

has a unique solution for any choice of Γ (thus, in particular, for the Γ defined in (24)). This equation can be split as follows

$$
\begin{align}
KX_0 + LHX_x - X_0\Phi &= 0 \tag{30} \\
\Phi\Pi + NHX_x - \Pi\Phi &= 0 \tag{31} \\
Z(0)X_z + T_x(0)HX_x - X_z\Phi &= 0 \tag{32} \\
GMX_0 + GT\Pi + P_z(0)X_z + (F+P_x(0))X_x - X_x\Phi &= G\Gamma . \tag{33}
\end{align}
$$

In view of the previous discussion, the second of these identities implies

$$
HX_x = 0 \tag{34}
$$

and, of course

$$\Phi\Pi = \Pi\Phi \,,$$

which is one of the two identities it is required to prove. Replacing (34) into (30) yields

$$KX_0 - X_0\Phi = 0$$

which in turn, since K has eigenvalues with negative real part by hypothesis, yields

$$X_0 = 0 \,. \tag{35}$$

Using (34) in (32) yields

$$Z(0)X_z - X_z\Phi = 0$$

which again implies

$$X_z = 0 \tag{36}$$

since $Z(0)$ has no eigenvalue with zero real part. Finally, replacing (35) and (36) into (33) yields

$$GT\Pi + (F + P_x(0))X_x - X_x\Phi = G\Gamma \,. \tag{37}$$

Using the identity (34), which expresses the fact that the first row of X_x is zero, and keeping in mind the special structures of F, G and $P_x(0)$ (which is lower triangular), the relation (37) yields

$$X_x = 0$$

i.e.

$$GT\Pi = G\Gamma \,.$$

which concludes the proof of (26).

The fact that the manifold $\mathcal{M}_c$ is invariant can be checked by direct substitution. ◁

Noting that the error e is zero on $\mathcal{M}_c$, the issue is now to choose $K, L, N, T, \alpha(\xi_0)$ in (25) so that $\mathcal{M}_c$ is semiglobally attractive. This issue will be addressed in the next section.

5 Semiglobal robust regulation

In the new coordinates

$$\begin{aligned}
\tilde{\xi}_1 &= \xi_1 - \Pi \tau^{\mathrm{a}}(w, \mu) \\
\tilde{z} &= z - \zeta(w, \mu) \\
\tilde{x} &= x - \pi^{\mathrm{a}}(w, \mu) ,
\end{aligned}$$

the closed loop system (23)-(25), driven by the exosystem (2), is described by equations of the form

$$\begin{aligned}
\dot{\xi}_0 &= K\xi_0 + LH\tilde{x} \\
\dot{\tilde{\xi}}_1 &= \Phi\tilde{\xi}_1 + NH\tilde{x} \\
\dot{\tilde{z}} &= Z(\mu)\tilde{z} + \tilde{p}_0(H\tilde{x}, w, \mu) \\
\dot{\tilde{x}} &= F\tilde{x} + G(\alpha(\xi_0) + T\tilde{\xi}_1) + \tilde{P}(\tilde{z}, \tilde{x}, w, \mu) \\
\dot{w} &= Sw
\end{aligned} \tag{38}$$

and, by construction,

$$\begin{aligned}
\tilde{p}_0(0, w, \mu) &= 0 \\
\tilde{P}(0, 0, w, \mu) &= 0
\end{aligned}$$

for every $(w, \mu) \in \mathbb{R}^d \times \mathcal{P}$. In these new coordinates, the invariant manifold $\mathcal{M}_c$ introduced in the previous section is the subspace where

$$\xi_0 = 0, \qquad \tilde{\xi}_1 = 0, \qquad \tilde{z} = 0, \qquad \tilde{x} = 0$$

and on this manifold the tracking error

$$e = \tilde{x}_1$$

is zero. Thus, output regulation is achieved if this manifold is attractive.

The issue is to render asymptotically stable the equilibrium point $(\xi_0, \tilde{\xi}_1, \tilde{z}, \tilde{x}) = (0, 0, 0, 0)$ of a time-varying system of the form

$$\begin{aligned}
\dot{\xi}_0 &= K\xi_0 + LH\tilde{x} \\
\dot{\tilde{\xi}}_1 &= \Phi\tilde{\xi}_1 + NH\tilde{x} \\
\dot{\tilde{z}} &= Z(\mu)\tilde{z} + \tilde{p}_0(H\tilde{x}, exp(St)w^\circ, \mu) \\
\dot{\tilde{x}} &= F\tilde{x} + G(\alpha(\xi_0) + T\tilde{\xi}_1) + \tilde{P}(\tilde{z}, \tilde{x}, exp(St)w^\circ, \mu) ,
\end{aligned} \tag{39}$$

in which w° represents the value at time $t = 0$ of the state of the exosystem. System (39) is an uncertain system because the actual value of μ as well as that of the initial state w° of the exosystem are not known. For consistency with the hypothesis on μ, we assume that the initial state w° of the exosystem ranges over an apriori known compact set $\mathcal{W} \in \mathbb{R}^d$. In other words, system (39) can be regarded as a system in which the terms $\tilde{p}_0(H\tilde{x}, exp(St)w^\circ, \mu)$ and $\tilde{P}(\tilde{z}, \tilde{x}, exp(St)w^\circ, \mu)$ are periodic (recall that the exosystem is assumed to be neutrally stable) functions of t, vanishing at $(\tilde{z}, \tilde{x}) = (0,0)$ for all $t \in \mathbb{R}$ and for all values of (w°, μ), unknown parameters ranging over a fixed compact set $\mathcal{W} \times \mathcal{P}$.

Robust asymptotic stability of the equilibrium point $(\xi_0, \tilde{\xi}_1, \tilde{z}, \tilde{x}) = (0,0,0,0)$ of (39), with a basin of attraction containing any arbitrarily large (but compact) set of initial conditions, can be achieved by appealing to a design philosophy which, for a similar robust stabilization problem, was proposed by Khalil in [18]. Recent progresses in the theory of robust stabilization due a number of authors, among which we refer to Kristic, Kanellakopoulos, Kokotovic [19], Teel and Praly [25] [26], and in the theory of input-to-state stability theory, introduced by Sontag (see e.g. [22]) and developed in a number of subsequent papers (see e.g. [23] [24]), are very useful to provide a streamlined presentation of this "semiglobal" robust stabilization philosophy.

The point of departure is the analysis of the state-feedback version of the problem in question. To this end, consider a control system described by equations of the form

$$
\begin{aligned}
\dot{\xi} &= \Phi\xi + NH\tilde{x} \\
\dot{\tilde{z}} &= Z(\mu)\tilde{z} + \tilde{p}_0(H\tilde{x}, exp(St)w^\circ, \mu) \\
\dot{\tilde{x}} &= F\tilde{x} + Gu + \tilde{P}(\tilde{z}, \tilde{x}, exp(St)w^\circ, \mu) \, ,
\end{aligned}
\tag{40}
$$

in which the input u is to be chosen, as a memoryless state-feedback, so as to achieve robust stability. In view of the particular structure of this system, it is clear that, if $\tilde{z}$ and $\tilde{x}$ were available for feedback (ξ is indeed available for feedback because the corresponding subsystem in (40) is part of the controller), global robust stabilization could be achieved via the powerful "backstepping" method developed by Kokotovic and coauthors. However, the feedback law needed in this case would be a *nonlinear* function of $\tilde{\xi}_1, \tilde{z}$ and $\tilde{x}$. On the contrary, if only *semiglobal* stablizability is sought, a simple *linear* feedback law can do the job. This possibility is particularly useful and can be exploited in the present setup, where linearity of the "internal model" (see previous section)

$$
\begin{aligned}
\dot{\xi}_1 &= \Phi\xi_1 + Ne \\
u &= T\xi_1
\end{aligned}
$$

has proven to be instrumental in establishing the existence of the *globally defined* invariant manifold $\mathcal{M}_c$.

In order to clarify this point, consider, as in [18], the (globally defined) coordinates transformation which replaces $\tilde{x}_1, \tilde{x}_2, \ldots, \tilde{x}_r$ with e and its first $r - 1$ derivatives with respect to time, and suppose the latter are available for feedback. Note that this will never be the case, even if the components of $\tilde{z}$ and $\tilde{x}$ were directly available, because the transformation in question would indeed depend on the unknown parameters w° and μ. However this apparent inconvenience can be overcome if, as suggested in [18], reasonable asymptotic estimates the first $r - 1$ derivatives of e can be generated. Using $\eta_1, \eta_2, \ldots, \eta_r$ to denote these new coordinates, we have (in view of "triangular" dependence of components of $\tilde{P}(\tilde{z}, \tilde{x}, exp(St)w^\circ, \mu)$ on the components of $\tilde{x}$)

$$
\begin{aligned}
\eta_1 &= \tilde{x}_1 \\
\eta_2 &= \tilde{x}_2 + \tilde{p}_1(\tilde{z}, \tilde{x}_1, exp(St)w^\circ, \mu) \\
&\cdots \\
\eta_r &= \tilde{x}_r + \tilde{p}_{r-1}(\tilde{z}, \tilde{x}_1, \ldots, \tilde{x}_{r-1}, exp(St)w^\circ, \mu) \, .
\end{aligned}
$$

Having changed the coordinates in this way, the control system (40) reduces to a system of the form

$$
\begin{aligned}
\dot{\xi} &= \Phi\xi + NH\eta \\
\dot{\tilde{z}} &= Z(\mu)\tilde{z} + p(H\eta, t, \theta) \\
\dot{\eta} &= F\eta + G(u + q(\tilde{z}, \eta, t, \theta)) \, ,
\end{aligned}
\tag{41}
$$

in which $p(H\eta, t, \theta)$ and $q(\tilde{z}, \eta, t, \theta)$ are periodic functions of t, vanishing at $(\tilde{z}, \eta) = (0, 0)$ for all t and θ, and in which $\theta = (w^\circ, \mu)$ is a $(d + p)$-tuple of unknown parameters, ranging over a fixed compact set Θ. Note also that, setting

$$
x_1 = \tilde{z}, \qquad x_2 = \begin{pmatrix} \xi \\ \eta \end{pmatrix}, \qquad A = \begin{pmatrix} \Phi & NH \\ 0 & F \end{pmatrix}, \qquad B = \begin{pmatrix} 0 \\ G \end{pmatrix}
$$

the system under consideration can be further simplified to

$$
\begin{aligned}
\dot{x}_1 &= Z(\mu)x_1 + \bar{p}(x_2, t, \theta) \\
\dot{x}_2 &= Ax_2 + B(u + \bar{q}(x_1, x_2, t, \theta)) \, ,
\end{aligned}
\tag{42}
$$

where $\bar{p}(x_1, t, \theta)$ and $\bar{q}(x_1, x_2, t, \theta)$ are periodic functions of t, vanishing at $(x_1, x_2) = (0, 0)$ for all t and $\theta \in \Theta$. If N is such that (Φ, N) is a controllable pair, also the pair (A, B) is controllable. Moreover, since all the entries on the last row of F are zero and the only nonzero entry in G is the last one (equal to 1), the vector x_2 and the matrices A and B can also be re-partitioned as

$$x_2 = \begin{pmatrix} x_{21} \\ x_{22} \end{pmatrix}, \quad A = \begin{pmatrix} A_{11} & A_{12} \\ 0 & 0 \end{pmatrix}, \quad B = \begin{pmatrix} 0 \\ 1 \end{pmatrix},$$

where $\dim(x_{21}) = q+r-1$ and $\dim(x_{22}) = 1$. By construction, (A_{11}, A_{12}) is a controllable pair.

The design procedure which will be illustrated relies upon the following hypothesis on (42).

Assumption A There exists a positive definite proper smooth function $V_1(x_1)$ satisfying

$$\frac{\partial V_1}{\partial x_1}(Z(\mu)x_1 + \bar{p}(x_2, t, \theta)) \leq -\alpha(V_1(x_1)) + c|x_2|^2 \qquad (43)$$

for all x_1, x_2, t and all $\theta \in \Theta$, where $\alpha(r)$ is a $\mathcal{K}_\infty$ function, which is also assumed to be continuously differentiable, and $c > 0$.
Moreover,

$$|\bar{q}(x_1, x_2, t, \theta)|^2 \leq \alpha(V_1(x_1)) + \gamma(|x_2|)$$

for all x_1, x_2, t and all $\theta \in \Theta$, where $\gamma(r)$ is a $\mathcal{K}_\infty$ function, which is also assumed to be continuously differentiable at $r = 0$.◁

Using backstepping techniques the following results can be proven.

Proposition 2 *Choose any matrix K_1 such that the eigenvalues of $A_{11}+ A_{12}K_1$ have negative real part, and let P be the unique (positive definite) solution of the Lyapunov equation*

$$(A_{11} + A_{12}K_1)^T P + P(A_{11} + A_{12}K_1) = -I .$$

Consider the positive definite function

$$V_2(x_2) = x_{21}^T P x_{21} + (x_{22} - K_1 x_{21})^2 .$$

Suppose assumption A holds. Then, for any (arbitrarily small) number $b > 0$ and any (arbitrarily large) number $d > 0$ there is a matrix K_2 such that

$$\frac{\partial V_2}{\partial x_2}((A+BK_2)x_2+Bv+B\bar{q}(x_1,x_2,t,\theta)) \leq -aV_2(x_2)+b\alpha(V_1(x_1))+bv^2$$

$$(44)$$

for all x_1, for all v, for all t, for all $\theta \in \Theta$ and for all x_2 such that $|x_2| \leq d$, where $a > 0$ is a number depending only on $V_2(x_2)$ (and not on b and d).

Clearly, system (42) can be viewed as the feedback interconnection of two subsystems one of which satisfies by hypothesis the input-to-state stability inequality (43) while the other one is such that an input-to-state stability inequality (see (44)) can be enforced by appropriate choice of a control law of the form

$$u = K_2 x_2 + v . \qquad (45)$$

Moreover, the "gain" of this feedback loop, which is proportional to the coefficient b in the inequality (44), can be rendered arbitrarily small by appropriate choice of the matrix K_2 in the control law (45). In other words, it is possible to fulfill the conditions of the small gain theorem (see e.g. [2] pages 335–339) so as to render input-to-state stable also the interconnection of (42) and (45), that is the system

$$\begin{aligned} \dot{x}_1 &= Z(\mu)x_1 + \bar{p}(x_2, t, \theta) \\ \dot{x}_2 &= (A + BK_2)x_2 + B\bar{q}(x_1, x_2, t, \theta) + Bv . \end{aligned} \qquad (46)$$

Remark. The details of the proof of this result, which are beyond the scope of this paper, can be figured out by appealing e.g. to the general methods for the analysis of interconnected input-to-state stable system presented in the above-mentioned reference [2]. As a sketch of what these methods consist, consider for instance the case in which $v = 0$ in (46) and semiglobal stability is sought. Let $c' > 0$ be any number such that $c|x_2|^2 \leq c'V_2(x_2)$. Then, from the previous Proposition we known that, for any choice of $b > 0$ and $d > 0$ there is K_2 such that,

$$\frac{\partial V_1}{\partial x_1}(Z(\mu)x_1 + \bar{p}(x_2, t, \theta)) \leq -\alpha(V_1(x_1)) + c'V_2(x_2)$$

$$\frac{\partial V_2}{\partial x_2}((A + BK_2)x_2 + B\bar{q}(x_1, x_2, t, \theta)) \leq -aV_2(x_2) + b\alpha(V_1(x_1)) \qquad (47)$$

for all x_1, for all x_2 such that $|x_2| \leq d$, for all t, for all $\theta \in \Theta$.
Set

$$\chi_1(r) = \alpha^{-1}(c'r), \qquad \chi_2(r) = \frac{b}{a}\alpha(r) .$$

The two inequalities in (47) show that, if $V_2(x_2) > \chi_2(V_1(x_1))$ then $V_2(x_2)$ is decreasing along the trajectories of the system, while if $V_1(x_1) >$

$\chi_1(V_2(x_2))$ then $V_1(x_1)$ is decreasing along the trajectories of the system. The corresponding regions in the (V_1, V_2) plane have a nonempty intersection if the (small gain) condition

$$\chi_1^{-1}(r) > \chi_2(r)$$

is fulfilled for all $r > 0$. This occurs, for instance, if

$$b = \frac{a}{4c'} \; .$$

Note now that, if b is chosen in this way, the function

$$\sigma(r) = 2\chi_2(r)$$

satisfies

$$\chi_1^{-1}(r) > \sigma(r) > \chi_2(r)$$

for all $r > 0$. Then, as suggested in [16], define

$$W(x_1, x_2) = \max\{V_2(x_2), \sigma(V_1(x_1))\} \; .$$

Note that this (positive definite and proper) function only depends on the function $V_1(x_1)$ introduced in the Assumption A and on the matrix K_1 chosen, once for all, in the way indicated at the beginning of the previous Proposition. In particular, $W(x_1, x_2)$ does not depend on the gain matrix K_2.

To prove semiglobal stabilizability, set

$$\Omega_c = \{(x_1, x_2) : W(x_1, x_2) \leq c\} \; .$$

Let S be any compact set of initial conditions in the (x_1, x_2) space and choose $\bar{c}$ be such that

$$S \subset \Omega_{\bar{c}} \; .$$

According to the value of $\bar{c}$ thus determined and to the value of b determined above, choose the matrix K_2 so as to render (44) fulfilled (for all $(x_1, x_2) \in \Omega_{\bar{c}+1}$). Then, using the property that $V_2(x_2)$ is decreasing whenever $V_2(x_2) > \chi_2(V_1(x_1))$ and $V_1(x_1)$ is decreasing whenever $V_1(x_1) > \chi_1(V_2(x_2))$, it is easy to see (as e.g. in [1]) that, for any $c \leq \bar{c}$, the set Ω_c is positively invariant for (46), with $v = 0$, and that any trajectory originating in $\Omega_{\bar{c}}$ converges to the origin as t tends to ∞.◁

The previous arguments show the existence, for any arbitrarily given compact set S of initial conditions, of a linear state feedback law yielding the desired property of robust asymptotic stability, with a basin of attraction which contains the set S. The problem is now to replace this law by an appropriate estimate, depending only on the variables which are actually accessible for feedback, the tracking error e and the state ξ_1 of the internal model. This problem can be successfully addressed via the method proposed by Esfandiari-Khalil in [5], and already used by Khalil in [18] to a similar robust stabilization purpose. Since the method in question is known (see also e.g. [25]), we limit ourselves to sketch only a few major details.

Suppose the state-feedback gain matrix K_2 has been fixed and set

$$K_2 = (T \quad M)$$

where two row vectors T and M on the right-hand side have dimension q (the dimension of ξ_1) and r (the dimension of x in (23)) respectively. Then, consider a control law of the form

$$
\begin{aligned}
\dot{\xi}_0 &= K\xi_0 + Le \\
\dot{\xi}_1 &= \Phi\xi_1 + Ne \\
u &= \psi(M\xi_0) + T\xi_1 ,
\end{aligned}
\tag{48}
$$

in which $\dim(\xi_0) = r$ and $\psi(\cdot)$ is a function satisfying $\psi(a) = a$ if $|a|$ is small. This is indeed a control law of the form (25) and therefore the result of Proposition 1 apply. If the hypotheses of this Proposition hold, then the corresponding closed-loop system has a globally defined invariant manifold on which the tracking error is zero. As shown at the beginning and throughout the section, to render the latter (semiglobally) attractive is equivalent to render semiglobally attractive the equilibrium $(\xi_0, \tilde{\xi}_1, \tilde{z}, \eta) = (0, 0, 0, 0)$ of an (uncertain) time-varying system of the form

$$
\begin{aligned}
\dot{\xi}_0 &= K\xi_0 + LH\eta \\
\dot{\tilde{\xi}}_1 &= \Phi\tilde{\xi}_1 + NH\eta \\
\dot{\tilde{z}} &= Z(\mu)\tilde{z} + p(H\eta, t, \theta) \\
\dot{\eta} &= F\eta + G(\psi(M\xi_0) + T\tilde{\xi}_1) + q(\tilde{z}, \eta, t, \theta) .
\end{aligned}
\tag{49}
$$

To this end, the matrix K, the vector L and the function $\psi(\cdot)$ must be determined (recall that N, M and T have already been fixed). The method of Esfandiari-Khalil consists in the following: choose for $\psi(\cdot)$ a saturation function, namely

$$\psi(a) = U_{\max}\text{sat}\left(\frac{a}{U_{\max}}\right)$$

where $U_{\max}$ is the maximum value of $M\eta$ on the set $\Omega_{\bar{c}+1}$ and choose for

$$\dot{\xi}_0 = K\xi_0 + Le$$

the structure of a high-speed observer, i.e. a system of the form

$$\dot{\xi}_0 = \begin{pmatrix} 0 & 1 & 0 & \cdots & 0 \\ 0 & 0 & 1 & \cdots & 0 \\ \cdot & \cdot & \cdot & \cdots & \cdot \\ 0 & 0 & 0 & \cdots & 1 \\ 0 & 0 & 0 & \cdots & 0 \end{pmatrix} \xi_0 + \begin{pmatrix} Ra_{r-1} \\ R^2 a_{r-2} \\ \cdot \\ R^{r-1} a_1 \\ R^r a_0 \end{pmatrix} (e - H\xi_0) \qquad (50)$$

in which $R > 0$ is a large number. As a matter of fact, it is possible to prove (as e.g. in [18]) that there exists a value $R^* > 0$ such that, if $R > R^*$ every trajectory of (49) with initial condition in a compact set of the form $\xi_0(0) \times S$ converges to the equilibrium as t tends to ∞.

Remark. An intuitive understanding of why the construction works can be obtained as follows. Consider the (R-dependent) change of coordinates

$$\chi = D(R)(\eta - \xi_0)$$

with

$$D(R) = \text{diag}\{R^{r-1}, \ldots, R, 1\}\,.$$

Moreover, set

$$v = \psi(M(\eta - D^{-1}(R)\chi)) - M\eta\,,$$

and note that, whilst χ can grow unbounded as R increases, the saturation effect introduced by $\psi(\cdot)$ prevents a similar phenomenon to occur for v.

Using the new coordinates thus defined and the coordinates x_1, x_2 introduced earlier, system (49) can be rewritten in the form

$$\begin{aligned} \dot{\chi} &= R\bar{K}\chi + G(K_2 x_2 + \bar{q}(x_1, x_2, t, \theta) + v) \\ \dot{x}_1 &= Z(\mu)x_1 + \bar{p}(x_2, t, \theta) \\ \dot{x}_2 &= (A + BK_2)x_2 + B\bar{q}(x_1, x_2, t, \theta) + Bv\,. \end{aligned} \qquad (51)$$

Suppose the eigenvalues of the matrix $\bar{K}$ (which can be assigned by appropriate choice of the a_i's in (50)) have negative real part, let Q be the solution of $\bar{K}^T Q + Q\bar{K} = -I$ and set $Q(\chi) = \chi^T Q\chi$. Then, it is immediate to check that for any (arbitrarily small) number $b' > 0$ there is a number $R > 0$ such that

$$\frac{\partial Q}{\partial \chi}(R\bar{K}\chi + Gw) \leq -a'Q(\chi) + b'w^2$$

for all χ and w, where a' is a number which depends only on $Q(x)$.

Taking advantage of this input-to-state stability property, of the input-to state stability property already shown for the remaining subsystem of the interconnection (49) and of the fact that v cannot not grow unbounded as R increases, it is possible to proceed to establish the desired convergence properties of the trajectories of (51).◁

Acknowledgments. The author wishes to thank Dr. Stefano Battilotti for many useful discussions. This work was supported in part by MURST and by NSF under grant ECS-9412340.

6 References

[1] S.Battilotti, Robust output feedback stabilization via small gain theorem, *Int. J. Robust Nonlinear Contr.*, to appear.

[2] J.M.Coron, L.Praly, A.Teel, Feedback stabilization of nonlinear systems: sufficient conditions and Lyapunov and input-output techniques, in *Trends in Control* (A.Isidori, ed.), Springer Verlag, pp.293–348, 1995.

[3] E.J.Davison, The robust control of a servomechanism problem for linear time-invariant multivariable systems, *IEEE Trans. Autom. Control*, **AC-21**: 25–34, 1976.

[4] F.Delli Priscoli, Robust tracking for polynomial plants, in *Proc. of 2nd European Control Conf.*: 369–373, Groeningen, The Netherlands, June 1993.

[5] F.Esfandiari, H.Khalil, Output feedback stabilization of fully linearizable systems, *Int. J. Control*, **56**: 1007–1037, 1992.

[6] M.Fliess, Finite dimensional observation spaces for nonlinear systems, in *Feedback control of linear and nonlinear systems* (D.Hinrichsen and A.Isidori, eds.): 73-77, Springer Verlag, 1982.

[7] B.A.Francis, The linear multivariable regulator problem, *SIAM J. Contr. Optimiz.*, **14**: 486–505, 1977.

[8] B.A.Francis, W.M.Wonham, The internal model principle of control theory, *Automatica*, **12**: 457–465, 1976.

[9] J.S.A.Hepburn, W.M.Wonham, Error feedback and internal model on differentiable manifolds, *IEEE Trans. Autom. Control*, **AC-29**: 397–403, 1984.

[10] J.S.A.Hepburn, W.M.Wonham, Structurally stable nonlinear regulation with step inputs, *Math. Systems Theory*, **17**: 319–333, 1984.

[11] J.Huang, C.F.Lin, On a robust nonlinear servomechanism problem, in *Proc. of 30th IEEE Conf. Decision and Control*: 2529–2530, Brighton, England, December 1991 (also on *IEEE Trans. Autom. Control*, **AC-39**: 1510–1513, 1994)

[12] J.Huang, W.J.Rugh, On the nonlinear multivariable servomechanism problem, *Automatica*, **26**: 963–972, 1990.

[13] J.Huang, W.J.Rugh, Stabilization on zero-error manifold and the nonlinear servomechanism problem, *IEEE Trans. Autom. Control*, **AC-37**: 1009–1013, 1992.

[14] A.Isidori, *Nonlinear control systems* (3rd. ed), Springer Verlag, 1995.

[15] A.Isidori, C.I.Byrnes, Output regulation of nonlinear systems, *IEEE Trans. Autom. Control*, **AC-35**: 131–140, 1990.

[16] Z.P. Jiang, I.M.Y. Mareels, Y. Wang, A Lyapunov formulation of nonlinear small gain theorem for interconnected ISS systems, *Automatica*, to appear.

[17] H.Khalil, Robustness servomechanism output feedback controllers for a class of feedback linearizable systems, oral presentation at *Workshop on nonlinear control systems*, St. Louis, Missouri, May 1992.

[18] H.Khalil, Robust servomechanism output feedback controllers for feedback linearizable systems, *Automatica*, **30**: 1587–1599, 1994.

[19] M.Krstic, I.Kanellakopoulos, P.Kokotovic, *Nonlinear Adaptive Control Design*, J.Wiley (New York), 1995.

[20] H.Knobloch, A.Isidori, D.Flockerzi, Topics in Control Theory, DMV-Seminar series, **22**, Birkhauser, 1993.

[21] R.Marino, P.Tomei, I.Kanellakopoulos, P.V.Kokotovic, Adaptive tracking for a class of feedback linearizable systems, *IEEE Trans. Autom. Control*, **AC-39**: 1314–1319, 1994.

[22] E.D.Sontag, Smooth stabilization implies coprime factorization, *IEEE Trans. Autom. Control*, **AC-34**: 435–443, 1989.

[23] E.D.Sontag, On the input-to-state stability property, *European J. Contr.*, **1**: 24–36, 1995.

[24] E.Sontag, Y.Wang, On characterizations of the input-to-state stability property, *Syst. Contr. Lett.* **24**: 351–359, 1995.

[25] A. Teel, L. Praly Global stabilizability and observability imply semiglobal stabilizability by output feedback, *Syst. Cont. Lett.*, **22**: 313-324, 1994.

[26] A. Teel, L. Praly Tools for semi-global stabilization by partial state and output feedback, *SIAM J. Contr. Optimiz.*, to appear.

3. Nonlinear Predictive Command Governors for Constrained Tracking

Edoardo Mosca*

Abstract

A method based on conceptual tools of predictive control is described for solving tracking problems wherein pointwise-in-time input and/or state inequality constraints are present. It consists of adding to a primal compensated system a nonlinear device called command governor (CG) whose action is based on the current state, set-point and prescribed constraints. The CG selects at any time a virtual sequence amongst a family of linearly parameterized command sequences by solving a convex constrained quadratic optimization problem, and feeds the primal system according to a receding horizon control philosophy. The overall system is proved to fulfill the constraints, be asymptotically stable, and exhibit an offset-free tracking behaviour, provided that an admissibility condition on the initial state is satisfied. Though the CG can be tailored for the application at hand by appropriately choosing the available design knobs, the required on-line computational load for the usual case of affine constraints is well tempered by the related relatively simple convex quadratic programming problem.

*Dipartimento di Sistemi ed Informatica, Università di Firenze, Via di S.Marta, 3 - 50139 Firenze, Italy, Tel. +39-55-4796258 - Fax. +39-55-4796363, E-mail: mosca@dsi.ing.unifi.it; http://www-dsi.ing.unifi.it/.

1   Introduction

In recent years there have been substantial theoretical advancements in the field of feedback control of dynamic systems with input and/or state-related constraints. For an account of pertinent results see [1, 2] which also include relevant references. Amongst the various approaches, the developments of this lecture are more akin to the predictive control methodology [3]-[7]. Predictive control, wherein the receding horizon control philosophy is used, selects the control action by possibly taking into account the future evolution of the reference. Such an evolution can be: known in advance, as in applications where repetitive tasks are executed, e.g. industrial robots; predicted, if a dynamic model for the reference is given; or planned in real time. This last instance is a peculiar and important potential feature of predictive control. In fact, taking into account the current value of both the state vector and the reference, a potential or *virtual* reference evolution can be designed on line so as to possibly make the related input and state responses fulfill pointwise-in-time inequality constraints. However, this mode of operation, whereby the reference is made state-dependent, introduces an extra feedback loop that complicates the stability analysis of the overall control system. This has been one of the reasons for which on-line reference design, though advocated for a long time as one of the key potential advantages of predictive control [6], [8]-[10], has received so far rare consideration in applications.

In most cases, predictive control computations amount to numerically solving on-line a high-dimensional convex quadratic programming problem. Though this can be tackled with existing software packages [11], it is a quite formidable computational burden if, as in predictive control, on-line solutions are required. In order to lighten computations, it is important to know how it is possible to borrow from predictive control the concept of on-line reference management so as to tackle constrained control problems by schemes requiring a lighter computational burden. The main goal of the present paper is to address this issue by laying down guidelines for synthesizing *command governors* (CG) based on predictive control ideas. A CG is a nonlinear device which is added to a primal compensated control system. The latter, in the absence of the CG, is designed so as to perform satisfactorily in the absence of constraints. Whenever necessary, the CG modifies the input to the primal control system so as to avoid violation of the constraints. Hence, the CG action is finalized to let the primal control system operate linearly within a wider dynamic range than that which would result with no CG. Preliminary studies along these lines have already appeared in [12]-[13], while more mature assessments of the related state of art can be found in [14] and [15]. For CGs approached from different perspectives see [16]-[21].

This lecture is organized as follows. Sect. 2 presents the problem formulation, and defines the CG based on the concept of a virtual command sequence. Some of the CG stability and performance features are also considered in Sect. 2. Sect. 3 discusses solvability aspects related to the CG optimization problem, and addresses the important practical issue of reducing to a fixed and off-line computable finite prediction-horizon the infinite time-interval over which the fulfilment of constraints has to be checked. Simulation examples are presented so as to exhibit the results achievable by the method.

2 Problem Formulation and Command Governor Design

Consider the following linear time-invariant system

$$\begin{cases} x(t+1) &=& \Phi x(t) + Gg(t) \\ y(t) &=& Hx(t) \\ c(t) &=& H_c x(t) + Dg(t) \end{cases} \tag{1}$$

In (1): $t \in \mathbb{Z}_+ := \{0, 1, \ldots\}$; $x(t) \in \mathbf{R}^n$ is the state vector; $g(t) \in \mathbf{R}^p$ the manipulable command input which, if no constraints were present, would essentially coincide with the output reference $r(t) \in \mathbf{R}^p$; $y(t) \in \mathbf{R}^p$ the output which is required to track $r(t)$; and $c(t) \in \mathbf{R}^{n_c}$ the constrained vector which has to fulfill the pointwise-in-time set-membership constraint

$$c(t) \in \mathcal{C}, \ \forall t \in \mathbb{Z}_+ \tag{2}$$

with $\mathcal{C} \subset \mathbf{R}^{n_c}$ a prescribed constraint set. The problem is to design a memoryless device [1]

$$g(t) := \underline{g}(x(t), r(t)) \tag{3}$$

---

[1]As discussed in next Remark 1, the results of this paper can be extended to the case when the computational delay is significant w.r.t. the sampling interval (unit time step).

in such a way that, under suitable conditions, the constraints (2) are fulfilled and possibly $y(t) \approx r(t)$. It is assumed that:

- Φ is a stability matrix, i.e. all its eigenvalues   (4)
  are in the open unit disk;
- System (1) is offset-free, i.e. $H(I - \Phi)^{-1}G = I_p$ (5)

One important instance of (1) consists of a linear plant under stabilizing linear state-feedback control. In this way, the system is compensated so as to satisfy stability and performance requirements, regardless of the prescribed constraints. In order to enforce the constraints, the CG (3) is added to the primal compensated system (1).

It is convenient to adopt the following notations for the equilibrium solution of (1) to a constant command $g(t) \equiv w$

$$\begin{cases} x_w & := & (I - \Phi)^{-1}Gw \\ y_w & := & Hx_w \\ c_w & := & H_c x_w + Dw = \left[H_c (I - \Phi)^{-1}G + D \right] w \end{cases} \qquad (6)$$

It is further assumed that:

- C is bounded;   (7)
- $C = \{c \in \mathbf{R}^{n_c} : q_j(c) \le 0, \ j \in \underline{n}_q\}$, with   (8)
  $\underline{n}_q := \{1, 2, ..., n_q\}$ and $q_j : \mathbf{R}^{n_c} \to \mathbf{R}$ continuous and convex;
- C has a non-empty interior.   (9)

Eq. (7) and (8) imply that C is compact and convex.

Consider a θ-parameterized family $\mathcal{V}_\Theta$ of sequences

$$\mathcal{V}_\Theta = \{v(\cdot, \theta) : \ \theta \in \Theta \subset \mathbf{R}^{n_\theta}\}, \quad v(\cdot, \theta) := \{v(k, \theta)\}_{k=0}^\infty \qquad (10)$$

with the property of closure w.r.t. left time-shifts, viz. $\forall \theta \in \Theta, \forall k \in \mathbb{Z}_+$, there exist $\bar{\theta} \in \Theta$ such that

$$v(k + 1, \theta) = v(k, \bar{\theta}) \qquad (11)$$

Suppose temporarily that $v(\cdot, \theta)$ is used as an input to (1) from the state $x(t)$ at time 0. The latter will be referred to as the event $(0, x(t))$. Assume that

$$c(\cdot, x(t), \theta) := \{c(k, x(t), \theta)\}_{k=0}^\infty \subset C \qquad (12)$$

In (12), $c(k, x(t), \theta)$ denotes the c-response at time k to $v(\cdot, \theta)$ from the event $(0, x(t))$. If the inclusion (12) is satisfied for some $\theta \in \Theta$, $(x(t), \theta)$ is said to be an *executable* pair, $x(t)$ *admissible*, and $v(\cdot, \theta)$ a *virtual command sequence* for the state $x(t)$. Notice that (11) ensures that

$$(x(t), \theta) \text{ is executable} \Longrightarrow \exists \bar{\theta} \in \Theta : (x(t+1), \bar{\theta}) \text{ is executable} \quad (13)$$

provided that $x(t+1) = \Phi x(t) + Gv(0, \theta)$. In fact, from (11) it follows that $c(k+1, x(t), \theta) = c(k, x(t+1), \bar{\theta})$. Then, any state is admissible along the trajectory corresponding to a virtual command sequence $v(\cdot, \theta)$. Consequently, no danger occurs of being trapped into a blind alley if (1) is driven by a virtual command sequence or its input switched from one to another virtual command sequence.

For reasons which will appear clear soon, it is convenient to introduce the following sets for a given $\delta > 0$:

$$\mathcal{C}_\delta := \{c \in \mathcal{C} : B_\delta(c) \subset \mathcal{C}\}, \quad (14)$$
$$B_\delta(c) := \{\bar{c} \in \mathbf{R}^{n_c} : \|c - \bar{c}\| \le \delta\}$$
$$\mathcal{W}_\delta := \{w \in \mathbf{R}^p : c_w \in \mathcal{C}_\delta\} \quad (15)$$

We shall assume that for a possibly vanishingly small $\delta > 0$

$$\mathcal{W}_\delta \text{ is non-empty} \quad (16)$$

From the foregoing definitions and (A.3), it follows that $\mathcal{W}_\delta$ is closed and convex.

There are various choices for the family $\mathcal{V}_\Theta$. Hereafter, we list three typical possibilities.

Case 1.

$$v(k, \theta) = \begin{cases} \theta_{k+1}, & k = 0, 1, \ldots, N-1 \\ w, & k = N, N = 1, \ldots \end{cases} \quad (17)$$
$$\theta := [\theta'_1 \theta'_2 \cdots \theta'_N w']' \in \Theta = \mathbf{R}^{pN} \times \mathcal{W}_\delta \quad (18)$$

Case 2.

$$v(k, \theta) = \gamma^k \mu + w, \quad (19)$$
$$\theta := [\mu' \ w']' \in \Theta = \mathbf{R}^p \times \mathcal{W}_\delta \quad (20)$$

Case 3.

$$v(k, \theta) = w, \quad \forall k \in \mathbf{Z}_+ \quad (21)$$
$$\theta := w \in \Theta = \mathcal{W}_\delta \quad (22)$$

Each of the three foregoing families has specific features. Sequences (17) have an arbitrary number of degrees-of-freedom which increases with N. On the other hand, the computational burden associated to any related CG would also increase with N. This makes the family $\mathcal{V}_\Theta$ corresponding to (17) very flexible in that N can be selected in accordance with a trade-off between performance (large N) and simplicity of operation (small N). The family $\mathcal{V}_\Theta$ of Case 3 is solely made of $\mathcal{W}_\delta$-valued constant sequences. It can be regarded as the simplex possible variant in Case 1: in fact, Case 3 is obtained by setting $N = 0$ in Case 1. Case 2 is intermediate between Case 1 and 3. Though it has only twice the number of degrees-of-freedom of case 3, the CG's based on the family $\mathcal{V}_\Theta$ of Case 2 can yield a very good tracking performance by a judicious choice of the γ value [14].

By the sake of simplicity, we shall only focus on Case 3, the family $\mathcal{V}_\Theta$ where $v(k, \theta)$ is as in (21). In such a case, instead of $c(k, x(t), \theta)$, we shall directly write $c(k, x(t), w)$.

We consider next the c-response $c(\cdot, x, w)$ of (1) to the command sequence (21). By straightforward manipulations we find

$$c(k) \;\; := \;\; c(k, x, w) = c_w + \tilde{c}(k) \tag{23}$$
$$\tilde{c}(k) \;\; := \;\; H_c \Phi^k [x - x_w] \tag{24}$$

In order to establish the existence of c-responses $c(\cdot, x, w) \subset C$, consider the special case $x = x_{\bar{w}}$ with $\bar{w} \in \mathcal{W}_\delta$. By stability of (1), there are two positive reals M and λ, $\lambda \in [0, 1)$, such that for each $x \in \mathbf{R}^n$ one has $\|\Phi^k x\| \leq M \lambda^k \|x\|$, $\forall k \in \mathbb{Z}_+$. Then, there are w, $\|\bar{w} - w\| > 0$, such that $c(\cdot, x, w) \subset C$. In fact, the following inequality holds for all $k \in \mathbb{Z}_+$

$$\|\tilde{c}(k)\| \leq \bar{\sigma}(H_c) M \|x_{\bar{w}} - x_w\| \tag{25}$$

with $\bar{\sigma}(H_c)$ the maximum singular value of H_c. Recalling that $x_w = (I - \Phi)^{-1} G w$, from (25) it follows that $\|\tilde{c}(k)\| \leq \delta$, $\forall k \in \mathbb{Z}_+$, provided that $\|x_{\bar{w}} - x_w\| \leq \delta/[\bar{\sigma}(H_c) M]$, or $\|\bar{w} - w\| \leq \delta(1 - \lambda)/[\bar{\sigma}(H_c)\bar{\sigma}(G)M^2]$. The foregoing analysis holds true if the initial state $x_{\bar{w}}$ is additively perturbed by $\tilde{x}$, $0 < \|\tilde{x}\| \leq \varepsilon$, with ε sufficiently small. In this case, the perturbed constrained vector $c(k)$ is such that $c(k) - c_w = H_c \Phi^k \tilde{x} + \tilde{c}(k)$. The condition $\|c(k) - c_w\| \leq \delta$, $\forall k \in \mathbb{Z}_+$, can be ensured, e.g., by requiring that $\|x_{\bar{w}} - x_w\| \leq \frac{1}{2}\delta/[\bar{\sigma}(H_c)M]$, and $\|\tilde{x}\| \leq \frac{1}{2}\delta/[\bar{\sigma}(H_c)M]$. The conclusion is that starting sufficiently close to an equilibrium state $x_{\bar{w}}$, $\bar{w} \in \mathcal{W}_\delta$, in a finite time one can arrive as close as desired to any state x_w, $w \in \mathcal{W}_\delta$, at a nonzero, though possibly small, distance from

$x_{\bar{w}}$. Then, we can move out from any admissible state $x(0)$ to reach asymptotically x_w, any $w \in \mathcal{W}_\delta$, by concatenating a finite number of virtual command sequences by switching from one to another, the last switching taking place at a finite, though possibly large, time. This result, which, by adopting the terminology of [22] will be referred to as a viability property, is summarized in the following Proposition 3.

Proposition 3 *(Viability property) Consider the system (1) along with the family of command sequences (21). Let the assumptions (4), (5), (7)-(9) and (16) be fulfilled and the initial state $x(0)$ of (1) admissible. Then, there exists a concatenation of a finite number of virtual constant command sequences $v(\cdot, w_i) = w_i$, $w_i \in \mathcal{W}_\delta$, with finite switching times, capable of asymptotically driving the system state from $x(0)$ to x_w, any $w \in \mathcal{W}_\delta$.*

Hereafter, we shall address the problem of how to select appropriate virtual constant command sequences, and when to switch from one to another. To this end, consider the quadratic selection index

$$J(r(t), w) := \|w - r(t)\|_{\Psi_w}^2 \tag{26}$$

where $\|w\|_{\Psi}^2 := w'\Psi w$, and $\Psi_w = \Psi_w' > 0$. Let $\mathcal{V}(x)$ be the set of all $w \in \mathcal{W}_\delta$ such that (x, w) is executable

$$\mathcal{V}(x) := \{w \in \mathcal{W}_\delta : c(\cdot, x, w) \subset \mathcal{C}\} \tag{27}$$

Assume that for every $t \in \mathbb{Z}_+$ $\mathcal{V}(x(t))$ is non-empty, closed and convex. This implies that the following minimizer exists unique

$$\begin{aligned}
w(t) &:= \arg \min_{w \in \mathcal{W}_\delta} \{J(r(t), w) : c(\cdot, x(t), w) \subset \mathcal{C}\} \\
&= \arg \min_{w \in \mathcal{V}(x(t))} J(r(t), w)
\end{aligned} \tag{28}$$

Proposition 3 ensures that $\mathcal{V}(x(t))$ non-empty implies $\mathcal{V}(x(t+1))$ non-empty if $(x(t), w)$ is executable and $x(t+1) = \Phi x(t) + Gw$. Further, the concatenation mechanism embedded in the viability property of Proposition 3 naturally suggests that we can select the actual CG action according to the following:

$$g(t) = w(t) \tag{29}$$

Remark. If the computational delay is not negligible w.r.t. the sampling interval, we can modify (28) as follows

$$w((i+1)\tau) = \arg \min_{w \in \mathcal{V}(x(i\tau))} J(r(i\tau), w)$$

$i \in \mathbb{Z}_+$, and set for $k = 0, 1, \ldots, \tau - 1$

$$g((i+1)\tau + k) = w((i+1)\tau)$$

This amounts to using an "open-loop" constant command sequence over intervals made up by τ steps. The results of this lecture can be extended to cover this case. ◁

Remark. As elaborated in some detail in next Example 2, the weighting matrix Ψ_w can be made $r(t)$-dependent so as to force the direction of the selected vector $w(t)$ to be as close as possible to that of $r(t)$, compatibly with the constraints. This can be a qualitatively important requirement in some MIMO applications. ◁

We defer the proof that $\mathcal{V}(x(t))$ is closed and convex to Sect. 3. A question we wish to address now is whether the foregoing CG yields an overall stable offset-free control system. Assume that the reference is kept constant, $r(t) \equiv r$ for all $t \geq t^*$, and $\mathcal{V}(x(t))$ is non-empty, closed and convex at each $t \in \mathbb{Z}_+$. Consider the following candidate Lyapunov function

$$V(t) := J(r, w(t)) = \|w(t) - r\|_{\Psi_w}^2 \tag{30}$$

If $x(t+1) = \Phi x(t) + Gw(t)$, it follows that $V(t) \geq V(t+1)$. In fact, $(x(t+1), w(t))$ is executable, but $w(t)$ need not be the minimizer of $J(r, w)$ at time $t+1$. It follows that along the trajectories of the system

$$V(t) - V(t+1) \geq 0 \tag{31}$$

Hence, $V(t)$, being nonnegative monotonically non increasing, has a finite limit $V(\infty)$ as $t \to \infty$.

Lemma 1 *Consider the system (1) controlled by the CG (28)-(29). Assume that (4), (5), (7)-(9) and (16) are satisfied. Let $x(0)$ be admissible and $\mathcal{V}(x(t))$ closed and convex at each $t \in \mathbb{Z}_+$. Let $r(t) \equiv r, \forall t \geq t^* \in \mathbb{Z}_+$. Then,*

$$V(\infty) := \lim_{\tau \to \infty} V(\tau) \leq V(t+1) \leq V(t), \ \forall t \geq t^* \tag{32}$$

and the CG output exhibits asymptotically vanishing variations in that

$$\lim_{t \to \infty} [w(t+1) - w(t)] = 0_p \tag{33}$$

Further,

$$\lim_{t \to \infty} [x(t) - x_{w(t)}] = 0_n \tag{34}$$

where $x_{w(t)} := (I - \Phi)^{-1} Gw(t)$.

Proof. At time $t+1$, both $w(t)$ and $w(t+1)$ are virtual constant references for the state $x(t+1)$: $w(t), w(t+1) \in \mathcal{V}(x(t+1))$. Then, by convexity of $\mathcal{V}(x(t+1))$, it follows that $\forall\, \alpha \in [0,1]$

$$w_\alpha(t) \in \mathcal{V}(x(t+1))$$
$$w_\alpha(t) := (1-\alpha)w(t) + \alpha w(t+1).$$

We have also for every $t \geq t^*$

$$
\begin{aligned}
V^{1/2}(\infty) &\leq \|w_\alpha(t) - r\|_{\Psi_w} = \|(1-\alpha)[w(t) - r] + \alpha[w(t+1) - r]\|_{\Psi_w} \\
&\leq (1-\alpha)\|w(t) - r\|_{\Psi_w} + \alpha\|w(t+1) - r\|_{\Psi_w}
\end{aligned}
\tag{35}
$$

where the upper bound follows by triangle inequality, and the lower bound by the monotonic property in (32). Because the upper and lower bounds in (35) both converge as $t \to \infty$ to $V^{1/2}(\infty)$, we find that

$$\lim_{t\to\infty} \|w_\alpha - r\| = V(\infty), \quad \forall\, \alpha \in [0,1].
\tag{36}$$

Now

$$w_\alpha(t) - r = \alpha[w(t+1) - w(t)] + [w(t) - r].$$

Then, (33) follows from (32) and (36).

In order to prove (34), let $\delta x(t) := x(t) - x(t+1)$ and $\delta w(t) := w(t) - w(t+1)$. We have $\delta x(t+1) = \Phi \delta x(t) + G\delta w(t)$. Then, stability of Φ and (33) yield

$$\lim_{t\to\infty} [x(t+1) - x(t)] = 0_n.$$

Further, let $\widehat{x}(t) = x(t) - x_{w(t)}$. We have,

$$\widehat{x}(t+1) = \Phi\widehat{x}(t) - G\delta w(t+1) - \Phi[x_{w(t+1)} - x_{w(t)}].$$

Then, (34) follows from (33) and stibility of Φ. $\qquad\square$

We are now ready to prove that, under the conditions stated after (29), the output of the system controlled by the CG converges to the best possible approximation to the reference.

Proposition 4 *(Finite stopping time)* *Under the same assumptions as in Lemma 1, the prescribed constraints are satisfied at every $t \in \mathbb{Z}_+$, and there is a time t_*, $t_* \geq t^*$, such that for all $t \geq t_*$*

$$w(t) = w_r := \arg\min_{w\in\mathcal{W}_\delta} \|w - r\|^2_{\Psi_w}
\tag{37}$$

and

$$\lim_{t \to \infty} y(t) = w_r \tag{38}$$

Proof. Assume that no such a t_* exist. Then, there are time instants t, $t = t(\epsilon)$ large enough such that

$$w(t) \neq w_r$$

and, by (34),

$$x(t) = x_{w(t)} + \tilde{x}(t)$$

with, by (34), $\|\tilde{x}(t)\| \leq \epsilon$, for any $\epsilon > 0$. Hence, by the viability property of Proposition 3, there are $t = t(\epsilon)$, $\epsilon = \epsilon(\bar{\alpha})$, with $\bar{\alpha} > 0$, such that

$$w(\alpha, t) := (1 - \alpha)w(t) + \alpha w_r \in \mathcal{V}(x(t))$$

for all $\alpha \in [0, \bar{\alpha}]$. A contradiction is obtained, if we can show that there are $\alpha \in (0, \alpha]$ for which

$$\|w(\alpha, t) - r\|_{\Psi_w} < \|w(t) - r\|_{\Psi_w} \tag{39}$$

Now it is easy to check that strict inequality in (39) holds provided that

$$0 < \alpha \leq 2 < 2\left(1 + \frac{[w_r - r]'\Psi_w[w(t) - w_r]}{\|w(t) - w_r\|_{\Psi_w}^2}\right)$$

where the rightmost inequality holds because of convexity of $\mathcal{W}_\delta$. □

3   Solvability and Computability

It remains to find existence conditions for the minimizer (28). Further, even if solvability is guaranteed, (28) embodies an infinite number of constraints. For practical implementation, we must find out if and how these constraints can be reduced to a finite number of constraints whose time locations be determinable off-line. To this end, it is convenient to introduce some extra notation. We express the response of (1) from an event $(0, x)$ to the constant command sequence (21) as follows

$$\begin{cases} z(k+1) &= Az(k), \text{ with } z(0) = \begin{bmatrix} x \\ w \end{bmatrix} \in \mathbf{R}^n \times \mathcal{W}_\delta, \\ c(k) &:= c(k, x, w) \\ &= E_c z(k) \end{cases} \tag{40}$$

where

$$A = \begin{bmatrix} \Phi & G \\ 0_{p\times n} & I_p \end{bmatrix}, \quad E_c = [H_c \ \ D] \tag{41}$$

For $i \in \mathbb{Z}_1 := \{1, 2, 3, ...\}$, consider the following sets

$$\mathcal{Z}_i := \{z \in \mathbf{R}^n \times \mathcal{W}_\delta : q_j(E_c A^{k-1} z) \leq 0, \ j \in \underline{n}_q, \ k \in \underline{i}\}. \tag{42}$$

$$\mathcal{Z} := \bigcap_{i=0}^{\infty} \mathcal{Z}_i \tag{43}$$

$\mathcal{Z}_i$ are the sets of initial states z with $w \in \mathcal{W}_\delta$ which give rise to evolutions fulfilling the constraints over the first i-th time steps $k = 0, 1, ..., i-1$, while $\mathcal{Z}$ is the set of all executable pairs (x, w). $\mathcal{Z}_{i+1} \subset \mathcal{Z}_i$, $\forall i \in \mathbb{Z}_1$, and under (A.2), all $\mathcal{Z}_i$'s, and hence $\mathcal{Z}$, are closed and convex. Moreover, by the viability property of Proposition 3 $\mathcal{Z}$ is non-empty.

Lemma 2

$$\mathcal{Z}_i = \mathcal{Z}_{i+1} \implies \mathcal{Z}_i = \mathcal{Z}.$$

Proof. Assume that $\mathcal{Z}_{i+1} = \mathcal{Z}_i$ and $[x' \ w']' \in \mathcal{Z}_i$. Then $c(k-1, x, w) \in C$, $\forall k \in \underline{i}$, implies that $c(k-1, x, w) \in C$, $\forall k \in \underline{i+1}$. By (11) one has that $c(i, x, w) = c(i-1, x_1, w)$ where $x_1 = \Phi x + Gw$. Then, $[x_1' \ w']' \in \mathcal{Z}_i$ and $[x' \ w']' \in \mathcal{Z}_{i+1}$. Similarly, one finds that $c(i, x_1, w) = c(i-1, x_2, w)$ with $x_2 = \Phi x_1 + Gw$. Then, $[x_2' \ w']' \in \mathcal{Z}_i$ and $[x' \ w']' \in \mathcal{Z}_{i+2}$. Thus, by induction, it is proved that $[x' \ w']' \in \mathcal{Z}_{i+l}$, $\forall l \in \mathbb{Z}_1$. The latter implies that $\mathcal{Z}_i \subset \mathcal{Z}_{i+l}$. Because $\mathcal{Z}_i \supset \mathcal{Z}_{i+l}$, it follows that $\mathcal{Z}_i = \mathcal{Z}_{i+l}$. Finally, $\mathcal{Z} = \bigcap_{k=0}^{\infty} \mathcal{Z}_k = \bigcap_{k=0}^{i} \mathcal{Z}_k = \mathcal{Z}_i$. $\square$

Consider next the "slice" of $\mathcal{Z}$ along x

$$\mathcal{V}(x) := \{w \in \mathcal{W}_\delta : \begin{bmatrix} x \\ w \end{bmatrix} \in \mathcal{Z}\}. \tag{44}$$

If x is admissible for some $w \in \mathcal{W}_\delta$, $\mathcal{V}(x)$ is non-empty. In addition, it is closed being the intersection of two closed sets, $\mathcal{V}(x) = \mathcal{Z} \cap \{\{x\} \times \mathcal{W}_\delta\}$. $\mathcal{V}(x)$ is also convex because the "slicer" operator is convexity-preserving. Then, existence and uniqueness of the minimizer (28) follows, provided that the initial state of (1) be admissible.

Proposition 5 *Let (4), (5), (7)-(9) and (16) be fulfilled and $(x(0), w)$ executable for some $w \in \mathcal{W}_\delta$. Thus, the optimization problem (28) is equivalent to the following convex constrained optimization problem*

$$w(t) := \arg \min_{w \in \mathcal{V}(x(t))} J(r(t), w), \; \forall t \in \mathbb{Z}_+ \tag{45}$$

This is uniquely solvable at each $t \in \mathbb{Z}_+$, being $\mathcal{V}(x(t))$ non-empty, closed and convex.

Proof. The viability property of Proposition 3 ensures that $\mathcal{V}(x(t))$ is non-empty. Existence and uniqueness of $w(t)$ follow because J is quadratic in w, and $\mathcal{V}(x(t))$ is also closed and convex. $\qquad\square$

Practical implementation of the CG requires an effective way to solve the optimization problem (45). Notice in fact that there might be no algorithmic procedure capable of computing the exact minimizer, unless $\mathcal{Z}$ is finitely determinable. In what follows, we shall show that only a finite number of pointwise-in-time constraints suffices to determine $\mathcal{Z}$. To this end, let (A_o, E_{co}), with $A_o \in \mathbf{R}^{n_o \times n_o}$, $n_o \leq n + p$, be an observable subsystem obtained via a canonical observability decomposition of (A, E_c). Then

$$c(k) = E_{c_o} A_o^k z_o(0) \tag{46}$$

with $z_o = P_o z$, P_o defined by the observability decomposition. Consequently, define the following sets

$$\mathcal{Z}_i^o := \{P_o z \in \mathbf{R}^{n_o} : z \in \mathcal{Z}_i\}, \quad \mathcal{Z}^o := \bigcap_{i=0}^{\infty} \mathcal{Z}_i^o \tag{47}$$

It is easy to see that $\mathcal{Z}_i^o$ and $\mathcal{Z}^o$ own the same properties shown to hold for $\mathcal{Z}_i$ and, respectively, $\mathcal{Z}$. In particular, they are non-empty, closed and convex. Moreover, the following result holds.

Proposition 6 *Let (4), (5), (7)-(9) and (16) be fulfilled. Then, $\mathcal{Z}_i^o$, $\forall i \geq n_o$ is non-empty, compact and convex. Moreover, there exists an integer $i_o \geq n_o$ such that $\mathcal{Z}_{i_o} = \mathcal{Z}$.*

Proof. Let $z_o \in \mathcal{Z}_{n_o}^o$. Because (A_o, E_{c_o}) is an observable pair, the related observability matrix $\Theta := \left[E_{c_o}' | (E_{c_o} A_o)' | ... | (E_{c_o} A_o^{n_o-1})' \right]'$ has full column rank. Hence $\Theta'\Theta$ is a nonsingular matrix, and

$$(\Theta'\Theta)z_o = \Theta'R, \text{ with } R = \left(\begin{bmatrix} c(0,x,w) \\ \vdots \\ c(n_o-1,x,w) \end{bmatrix}\right)$$

It follows that, $z_o = (\Theta'\Theta)^{-1}\Theta'R$. Then, being C bounded, $\mathcal{Z}^o_{n_o+l}$, $\forall l \in \mathbb{Z}_+$, is bounded as well, because $\mathcal{Z}^o_{n_o+l} \subset \mathcal{Z}^o_{n_o}$, $\forall l \in \mathbb{Z}_+$. In order to show that $\mathcal{Z}$ is finitely determinable, note that $\lim_{i\to\infty} c(i,x,w) = c_w$. Now

$$\begin{aligned} c(i,x,w) - c_w &= E_c M^i(z - z_w) \\ &= E_{co}M_o^i(z_o - z_{wo}) \end{aligned}$$

where $M = \begin{bmatrix} \Phi & G \\ 0_{p \times n} & 0_{p \times p} \end{bmatrix}$, $z_w = \begin{bmatrix} x_w \\ w \end{bmatrix}$, M_o is obtained from M in the same way as A_o from A, and $z_o = P_o z$ and $z_{wo} = P_o z_w$. Then,

$$\|c(i,x,w) - c_w\| \le \bar{\sigma}\left(E_{co}M_o^i\right)(\|z_o\| + \|z_{wo}\|)$$

Because $z_{wo} \in \mathcal{Z}^o$, $\|z_o\| + \|z_{wo}\|$ is bounded for all $z_o \in \mathcal{Z}^o$. Therefore, the existence of an integer i_o such that

$$i \ge i_o \implies \|c(i,x,w) - c_w\| \le \delta, \quad \forall z \in \mathcal{Z}$$

follows from asymptotic stability of M. $\qquad\qquad\square$

It follows that $\mathcal{Z}^o$, and hence $\mathcal{Z}$ as well, is finitely determinable, that is it suffices to check the constraints over the initial i_o time steps in order to ensure constraint fulfillment over $\mathbb{Z}_+$. Consequently, problem (45) is equivalent to the following finite dimensional convex constrained optimization problem at each $t \in \mathbb{Z}_+$:

$$\begin{aligned} w(t) \quad := \quad & \arg\min_{w \in W_s} J(r(t), w) \\ & \text{subject to } q_j(c(i-1, x(t), w)) \le 0, \ j \in \underline{n}_q, \ i \in \underline{i}_o \end{aligned} \tag{48}$$

The Gilbert and Tan algorithm [18] can be adapted to the present case to find $i_o = \min_{i \ge n_o}\{i \mid \mathcal{Z}^o_i = \mathcal{Z}^o\}$. To this end, let

$$\begin{aligned} G_i(j) \quad := \quad & \max_{w \in W_s}\{q_j(c(i,x,w))\}, \ j \in \underline{n}_q, i = 1, 2, \dots \\ & \text{subject to } q_j(c(k-1, x, w)) \le 0, \ j \in \underline{n}_q, \ k \in \underline{i} \end{aligned} \tag{49}$$

Then, i_o can be computed off-line via the following algorithm:

$$
\left.\begin{array}{ll}
1. & i \leftarrow n_o; \\
2. & \text{Solve } G_i(j), \ \forall j \in \underline{n}_q; \\
3. & \text{If } G_i(j) \leq 0, \ \forall j \in \underline{n}_q, \text{ let } i_o = i \text{ and stop}; \\
4. & \text{Otherwise } i \leftarrow i+1, \text{ and go to 2.}
\end{array}\right\} \qquad (50)
$$

Notice that step 2. in (50) is well posed because, according to Proposition 6, the implied maximization is carried out over a compact and convex set. In conclusion, we have found that our initial optimization problem having an infinite number of constraints is equivalent to a convex constrained optimization problem with a finite number of constraints.

Theorem 2 *Let (4), (5), (7)-(9) and (16) be fulfilled. Consider the system (1) with the CG (28)-(29), and let $x(0)$ be admissible. Then:*

  i. *The J-minimizer (45) uniquely exists at each $t \in \mathbb{Z}_+$ and can be obtained by solving a convex constrained optimization problem with inequality constraints $q_j(c(i-1, x(t), w)) \leq 0, \ j \in \underline{n}_q$, limited to a finite number i_o of time-steps, viz. $i = 1, ..., i_o$;*

 ii. *The integer i_o can be computed off-line via (50);*

iii. *The overall system satisfies the constraints, is asymptotically stable and off-set free, in that the conclusions of Proposition 4 hold.*

4   Simulation studies

The simulation results reported hereafter were obtained under Matlab 4.0 + Simulink 1.2 on a 486 DX2/66 personal computer, with no particular care of code optimization. The standard Matlab QP.M routine was used for quadratic optimization.

Example 1 Consider the following nonminimum-phase SISO system

$$
y(t) = \frac{-0.8935z + 1.0237}{z^2 - 1.5402z + 0.6703} g(t) \qquad (51)
$$

The unit step response of (51) is depicted in Fig. 1a (thin line). The task of the CG is to bound the output between -0.5 and 5. Accordingly, $c(t) = y(t)$ and $C = [-0.5, 5]$. Algorithm (50) takes 3.8s to give $i_o = 14$. The related constrained unit step response is shown in Fig. 1a

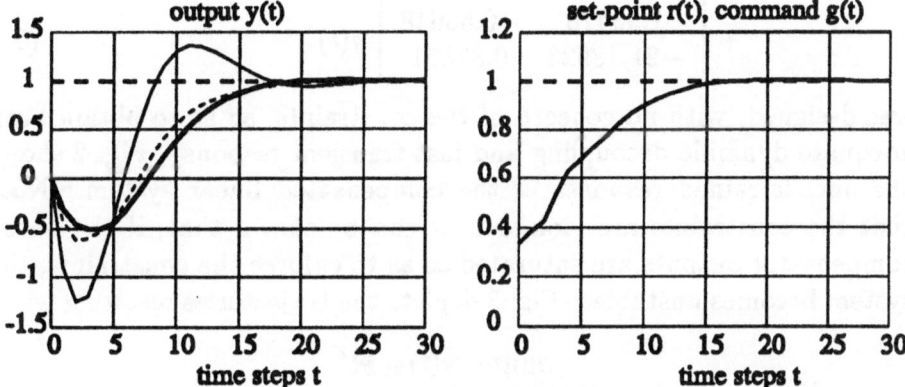

Figure 1: Example 1: (a) [left] Unit step response with no CG (thin line) and with CG ($\delta = 0.05$; thick line) for the nominal plant (51); Response with CG for the perturbed plant (52) (dashed line). (b) [right] Reference trajectory $r(t)$ (thick dashed line); Generated command trajectory $g(t)$ (thick line).

(thick line). This was computed in $0.1s$ per time step. Fig. 1b depicts the generated command trajectory $g(t)$ (thick line), and the reference trajectory $r(t)$ (thick dashed line).

In order to consider effects of model uncertainties, the same CG as the one designed for the nominal plant (51) was used with the perturbed plant

$$y(t) = \frac{-1.2517z + 1.4352}{z^2 - 1.4657z + 0.6492}u(t) \qquad (52)$$

Fig. 1a exhibits the related output response (thick dashed line). The prescribed lower bound is slightly violated.

Example 2 The CG is applied to the AFTI-16 aircraft modelled in continuous-time as in [17]. The elevator and flaperon angles are the two components u_i, $i = 1, 2$, of input u to the plant. They are subject to the physical constraints $|u_i| \leq 25°$, $i = 1, 2$. Then, $c = u$. The attack and the pitch angles are the two components of the output y. The task is to get zero offset for piecewise-constant references, while avoiding input saturations. The continuous-time model in [17] is sampled every $T_s = .05s$ and a zero-order hold is used at the input. The following linear compensator

$$u(t) = \begin{bmatrix} 0.00005 & 1.25601 & -0.17872 & 0.55620 \\ -0.00043 & 13.71101 & 4.06960 & -0.37350 \end{bmatrix} x(t) +$$

$$+ \begin{bmatrix} 1.93476 & -0.55618 \\ -21.18923 & 0.37351 \end{bmatrix} g(t) \tag{53}$$

was designed, with no concern of the constraints, so as to obtain both adequate dynamic decoupling and fast transient response. Fig. 2 shows the unconstrained response of the compensated linear system. Note that the constraints are violated. It can be shown that, if the linear compensator outputs are saturated so as to enforce the constraints, the system becomes unstable. Fig. 3 depicts the trajectories resulting when

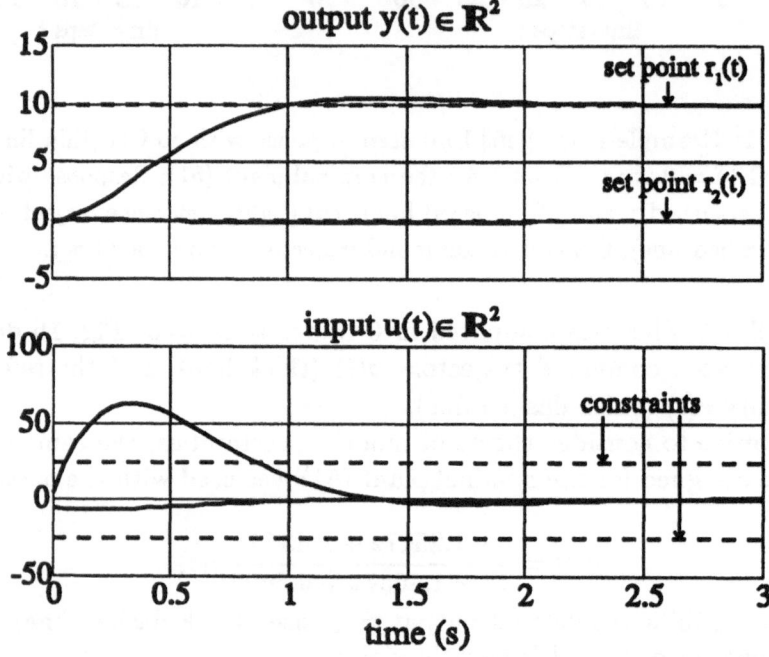

Figure 2: Example 2. Compensated AFTI-16 response with no CG.

the CG is activated so as to constrain the two plant inputs within the prescribed bounds. To this end, after some simulation analysis, we tuned the CG design knobs as follows: $\delta = 0.1$ and $\Psi_w = I_2$. Under these choices, algorithm (50) finds $i_o = 140$. Simulations were carried out with a computational time of 0.5s per step. Heuristically, it was found that, for the reference sequence of interest, indistinguishable results can be obtained with a constraint horizon equal to 5 in 0.1s per time step. Though these computational times exceed the sampling interval T_s, the simulation results indicate the performance which could be achieved by using faster processors with software specifically optimized for the application at hand. Because of vector optimization, the reference is

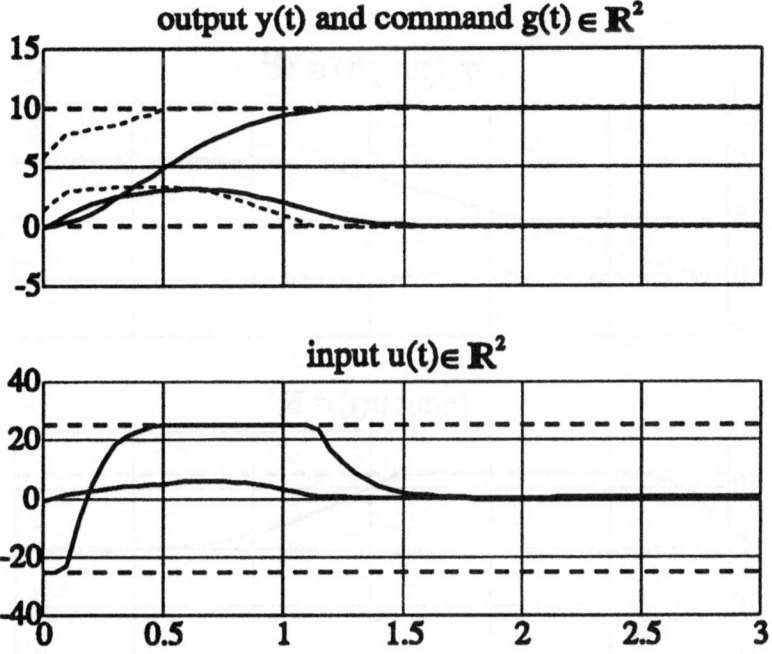

Figure 3: Example 2. Response with the CG: output $y(t)$ (solid line), command $g(t)$ (thin dashed line), and set-point $r(t)$ (thick dashed line).

filtered both in modulus and direction. This explains the crosscoupling between the two outputs. In order to let the direction of $g(t)$ be as close as possible to that of $r(t)$, Ψ_w was modified by penalizing at each time t also the component of $g(t)$ orthogonal to $r(t)$. This is accomplished by adding to Ψ_w the weighting matrix

$$100(I - \frac{r(t)r'(t)}{r'(t)r(t)}) = \begin{bmatrix} 100 & 0 \\ 0 & 0 \end{bmatrix} \tag{54}$$

The related trajectories, as depicted in Fig. 4, exhibit a reduced cross-coupling at the cost of longer settling times. Fig. 4 shows the performance of the system with the foregoing CG when the reference exhibits time-variations in such a way that transients take place also for non-equilibrium states.

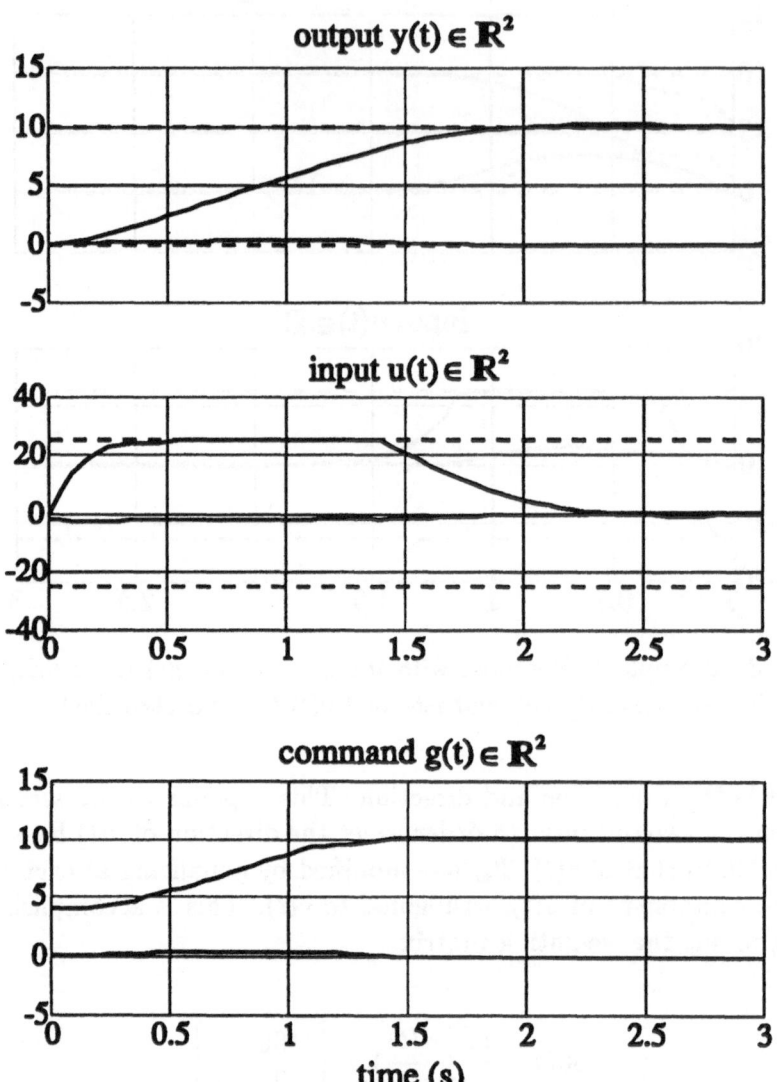

Figure 4: Example 2. Response with the CG penalizing the component of w orthogonal to $r(t)$.

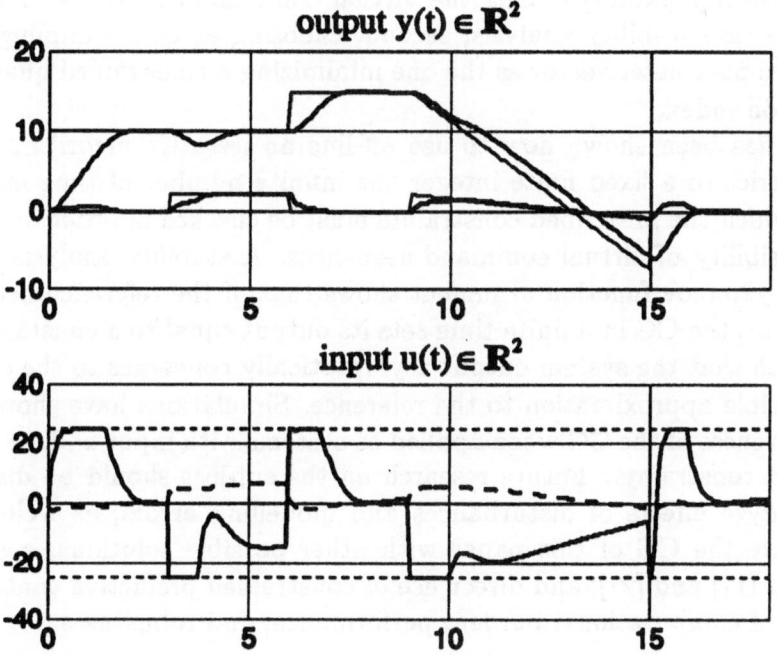

Figure 5: Example 2. Output $y(t)$ (thick line) and set-point trajectory $r(t)$ (thin line); input $u(t)$ (solid and dashed line)

5   Conclusions

The CG problem, viz. the one of on-line selecting a command input in such a way that a primal compensated control system can operate in a stable way with satisfactory tracking performance and no constraint violation, has been addressed by exploiting some ideas originating from predictive control. In this connection, the concept of a "virtual" command sequence is instrumental to synthesize CG's having the stated properties along with a moderate computational burden. This is achieved by: first, linearly parameterizing the virtual command sequences in such a way to ease stability analysis; second, choosing at each sampling time the free parameter vector as the one minimizing a constrained quadratic selection index.

It has been shown how to use off-line an iterative algorithm so as to restrict to a fixed finite integer the infinite number of time-instants over which the prescribed constraints must be checked in order to decide admissibility of virtual command sequences. A stability analysis based on a Lyapunov function argument shows that, if the reference becomes constant, the CG in a finite time sets its output equal to a constant vector such that the system output asymptotically converges to the closest admissible approximation to the reference. Simulations have shown the effectiveness of the CG when applied to systems with input and/or state-related constraints. Future research on the subject should be directed to analyze effects of disturbances and modelling errors, as well as to compare the CG of this paper with other possible solutions, e.g. the ones in [17] and [21], and direct use of constrained predictive control, in terms of computational burden, performance, and robustness.

6   References

[1] D. Q. Mayne and E. Polak, "Optimization based design and control", *Preprints 12th IFAC World Congress*, Vol. 3, pp. 129-138, Sydney, Jul. 1993.

[2] H.J. Sussmann, E.D. Sontag and Y. Yang, "A general result on the stabilization of linear systems using bounded controls", *IEEE Trans. Automat. Control*, Vol. 39, pp. 2411-2424, 1994.

[3] S. S. Keerthi and E. G. Gilbert, "Optimal infinite-horizon feedback control laws for a general class of constrained discrete-time systems:

stability and moving-horizon approximations", *J. Opt Theory and Applications*, Vol. 57, pp. 265-293, 1988.

[4] D. Q. Mayne and H. Michalska, "Receding horizon control of nonlinear systems", *IEEE Trans. Automat. Control*, Vol. 35, pp. 814-824, 1990.

[5] J. B. Rawlings and K. R. Muske, "The stability of constrained receding-horizon control", *IEEE Trans. Automat. Control*, Vol. 38, pp. 1512-1516, 1993.

[6] E. Mosca, *Optimal, Predictive, and Adaptive Control*, Prentice Hall, Englewood Cliffs, N. Y., 1995.

[7] D. W. Clarke, "Advances in Model-Based Predictive Control", *Advances in Model-Based Predictive Control*, Oxford University Press Inc., N. Y., pp. 3-21, 1994.

[8] J. M. Martín Sanchez, "A new solution to adaptive control", *Proc. IEEE*, Vol. 64, pp. 1209-1218, 1976.

[9] R. Soeterboek, *Predictive Control. A Unified Approach*, Prentice-Hall, Englwood Hill, N. J., 1992.

[10] J. Richalet, "Industrial applications of model based predictive control," *Automatica*, Vol. 29, pp. 1251-1274, 1993.

[11] S. P. Boyd and C. H. Barrat, *Linear Controller Design: Limits of Performance*, Prentice-Hall, Englewood Cliffs, N. J. , 1991.

[12] A. Bemporad and E. Mosca, "Constraint fulfilment in feedback control via predictive reference management," *Proc. 3rd IEEE Conf. on Control Applications*, pp. 1909-1914, Glasgow, U. K., 1994.

[13] A. Bemporad and E. Mosca, "Nonlinear predictive reference governor for constrained control systems", *Proc. 34th IEEE Conf. on Decision and Control*, pp. 1205-1210, New Orleans, Lousiana, U.S.A., 1995.

[14] A. Bemporad, A. Casavola and E. Mosca, "Nonlinear control of constrained linear systems via predictive reference management," submitted *IEEE Trans. Automat. Control*, Dec. 1995.

[15] A. Bemporad, A. Casavola and E. Mosca, "A nonlinear command governor for constrained control systems", *13th IFAC World Congress*, San Fransisco, California, U.S.A., 1996.

[16] P. Kapasouris, M. Athans and G. Stein, "Design of feedback control systems for stable plants with saturating actuators," *Proc. 27th IEEE Conf. on Decision and Control*, pp. 469-479, Austin, Texas, U.S.A., 1988.

[17] P. Kapasouris, M. Athans and G. Stein, "Design of feedback control systems for unstable plants with saturating actuators," *Proc. IFAC Symp. on Nonlinear Control System Design*, pp. 302-307, Pergamon Press, 1990.

[18] E. G. Gilbert and K. Tin Tan, "Linear systems with state and control constraints: the theory and applications of maximal output admissible sets," *IEEE Trans. Automat. Control*, Vol. 36, pp. 1008-1020, 1991.

[19] E. G. Gilbert, I. Kolmanovsky and K. Tin Tan, "Discrete-time reference governors and the nonlinear control of systems with state and control constraints", *Proc. 33rd IEEE Conf. on Decision and Control*, pp. 144-194, Lake Buena Vista, FL., U.S.A., 1994.

[20] T. J. Graettinger and B. H. Krogh, "On the computation of reference signal constraints for guaranteed tracking performance," *Automatica*, Vol. 28, pp. 1125-1141, 1992.

[21] E. G. Gilbert, I. Kolmanovsky and K. Tin Tan, "Discrete-time reference governors and the nonlinear control of systems with state and control constraints", *Int. Journal of Robust and Nonlinear Control*, Aug. 1995.

[22] J. P. Aubin, "*Viability Theory*", Birkhäuser, Boston, 1991.

4. Multivariable Regulation in Geometric Terms: Old and New Results

Giovanni Marro*

Abstract

The aim of this lecture is to present some well-known and some new results in multivariable regulation in a strictly geometric framework, each compared with the corresponding single-variable result approached with the standard transfer function techniques. It consists of three parts, each necessary for a clear presentation of the subsequent one: a selection of the basic tools, a survey of the solution of the asymptotically robust autonomous regulator problem with the internal model of the exosystem, and a presentation of the "steering along zeros technique" to obtain multivariable perfect tracking in the minimum-phase case or almost perfect tracking by using preaction in the nonminimum-phase case.

1 Introduction

The Geometric State-Space Theory is a collection of mathematical concepts developed to achieve a better and neater insight into the most salient features of multivariable linear dynamical systems in connection with compensator and regulator synthesis problems. It is based on the state space representation and is also very useful to easily link SISO and

*Dipartimento di Elettronica Informatica e Sistemistica (DEIS), Università di Bologna, Viale Risorgimento, 2, 40136 Bologna, Italia, Tel.: +39-51-6443046, Fax: +39-51-6443073, E-mail: gmarro@deis.unibo.it

MIMO systems and to clarify in quite a concise and elegant way some common properties that cannot be pointed out by the transform-based techniques usually adopted in the SISO case. Although the geometric tools are very simple and supported by exhaustive computational machinery, it is rather difficult to get a complete panorama of them, since their presentation in the literature by several authors over a period of more than 25 years is not uniform in style and has very often been covered in unnecessarily heavy mathematics.

At the end of the sixties (1969) Basile, Marro and Laschi published some results using geometric techniques in five papers, three in Italian and two in English, that analyzed the basic tools and presented solutions to some problems to which they could be profitably applied: disturbance rejection [6], unknown-input observability [5,35] and noninteraction [3]. The first of the two papers in English [4] presented definitions and properties of the major protagonists of this approach, that were called "controlled and conditioned invariants". At about the same time Wonham and Morse applied an algorithm similar to that for the maximal controlled invariant computation for the solution of noninteracting control problems [57]. In their paper only the algorithm was presented, while the name "invariant with respect to (A,B)", to be later transformed into "(A,B)-invariant" was only introduced by the same authors about two years afterwards [44].

However, the work of Wonham and Morse during the seventies, in many cases in cooperation with Francis and Pearson [26–28,54,56,58], was conclusive to investigate and propagate the use of the geometric techniques, in particular referring to multivariable tracking and regulation problems. Wonham's book [55] is still an exhaustive standard reference for this approach.

In the eighties major contributions are due to Willems, with the theory of almost controlled and almost conditioned invariant subspaces to deal with high-gain feedback problems [51, 52], Willems and Commault [53] and Schumacher [47], who contributed with a complete study of the regulation problem, including structural disturbance rejection. The use of self-bounded controlled and self-hidden conditioned invariant subspaces by Basile and Marro [9, 12, 13] allowed for a more direct presentation of these results. Introduction of the robust controlled invariant, also by Basile and Marro [10], opened the way to new applications, for instance the elimination of regulation transients in multivariable large-scale plants under parameter jumps or configuration changes, recently approached by Marro and Piazzi [39].

This lecture consists of three sections. The purpose of the first is to present, in the simplest possible terms, a selection of the basic tools of the geometric approach, strictly necessary to discuss multivariable tracking and regulation. The tools are: controlled and conditioned invariants, self-bounded controlled invariants, system invertibility, perfect output controllability, constrained output controllability (multivariable relative degree), and invariant zeros. Since this selection is strictly oriented to treatment of tracking and regulation, it does not include some other tools, like almost controlled and conditioned invariants and the robust controlled invariant. Then, referring to the classical disturbance rejection problem with stability, it is shown that the geometric solvability conditions usually consist of two parts: a structural condition and a stabilizability condition; the latter can be stated both in terms of self-bounded controlled invariants or in terms of invariant zeros, and can be relaxed if there is some preview of the signal to reject.

The second section deals with the central points of feedback regulation: inclusion of an internal model of the exosystem in the regulator loop and achievement of the closed-loop pole assignment: the SISO solution, that can be based on the Diophantine equation, is compared with the MIMO solution with the geometric techniques, and robustness of MIMO regulation, that requires as many replicas of the internal model as the number of regulated outputs, is briefly discussed.

The third section introduces perfect tracking as a further requirement for the previously derived asymptotic regulator and resorts to feedforward as a second-degree-of-freedom feature. A solution with the standard transfer function analysis for the SISO case and the corresponding geometric solution for the MIMO case are presented, in which a special feedforward unit, that reproduces an approximate replica of the reference function by switching an exosystem and suitably filtering the obtained signal, is proposed to obtain perfect or almost perfect tracking also in the nonminimum-phase case.

2 The Tools

In this section a brief review of the basic geometric tools is presented. By "basic" we understand:

1. strictly necessary to state a self-contained mathematical background for the theory of multivariable regulation and tracking pre-

sented in next sections;

2. strictly necessary to support the computational framework for the constructive numerical solution of all the presented problems.

They are: invariants, controlled invariants and self-bounded controlled invariants, conditioned invariants, system invertibility and functional output controllability, invariant zeros.

Let us note incidentally that the geometric tools are not only used for a neat presentation of multivariable control theory, but also provide a useful computational framework, based on few algorithms that extend some basic straightforward computations of the standard state-space approach to system theory (controllability, observability and pole assignment). Hence, the selection of the basic tools described in this section is also related to their impact on the computational framework.

The following notation is used. $\mathbf{R}$ stands for the field of real numbers, $\mathbf{C}$ for that of complex numbers, split into $\mathbf{C}_-$, $\mathbf{C}_0$ and $\mathbf{C}_+$ (left half-plane, imaginary axis and right half-plane). Sets, vector spaces and subspaces are denoted by script capitals like $\mathcal{X}$, $\mathcal{I}$, $\mathcal{V}$, etc.; since most of the geometric theory of dynamic system herein presented is developed in the vector space $\mathbf{R}^n$, we reserve the symbol $\mathcal{X}$ for the full space, i.e., we assume $\mathcal{X} := \mathbf{R}^n$. The orthogonal complement of any subspace $\mathcal{Y} \subseteq \mathcal{X}$ is denoted by $\mathcal{Y}^\perp$, matrices and linear maps by slanted capitals like A, B, etc., the image and the null space of the generic matrix or linear transformation A by $\mathrm{im}A$ and $\ker A$ respectively, the transpose of the generic real matrix A by A^T, the spectrum of A by $\sigma(A)$, the $n \times n$ identity matrix by I_n.

The algorithms of the geometric approach require computations involving subspaces. A generic subspace $\mathcal{X}_1 \subseteq \mathcal{X}$ is numerically defined with a *basis matrix*, i.e., a matrix X_1 having maximal rank such that $\mathcal{X}_1 = \mathrm{im}X_1$. The operations on subspaces that are required in the geometric-type algorithms are the sum $\mathcal{X}_1 + \mathcal{X}_2$, the orthogonal complementation $\mathcal{X}_1^\perp$, the intersection $\mathcal{X}_1 \cap \mathcal{X}_2 = (\mathcal{X}_1^\perp + \mathcal{X}_2^\perp)^\perp$, the direct linear mapping $A\mathcal{X}_1$ (where $A : \mathcal{X} \to \mathcal{X}$ denotes a linear map represented by the $n \times n$ matrix A with respect to the main basis of $\mathcal{X}$), the inverse linear mapping $A^{-1}\mathcal{X}_1 = (A^T \mathcal{X}_1^\perp)^\perp$.

The linear dynamic system herein considered as the reference to introduce the geometric tools is

$$\dot{x}(t) = A\,x(t) + B\,u(t), \quad x(0) = x_0 \tag{2.1}$$
$$y(t) = C\,x(t) \tag{2.2}$$

where $x \in \mathcal{X}$ ($= \mathbf{R}^n$), $u \in \mathcal{U}$ ($= \mathbf{R}^p$) and $y \in \mathcal{Y}$ ($= \mathbf{R}^q$) denote respectively the state, input and output. Let $\mathcal{B} := \mathrm{im} B$ and $\mathcal{C} := \ker C$. System (2.1, 2.2) is simply called the *triple* (A, B, C), while the sole differential equation (2.1), describing the state evolution caused by the input $u(t)$ with $x_0 = 0$, is called the *pair* (A, B), and the set consisting of equation (2.2) and equation (2.1) with only the first term on the right, describing the state evolution due to the initial state x_0, is called the *pair* (C, A). Let us also recall that the two basic algebraic feedback connection considered in synthesis procedures, that are state-to-input through a matrix F, that transforms the triple (A, B, C) into $(A+BF, B, C)$, and output-to-state through a matrix G, that produces $(A+GC, B, C)$, are called *state feedback* and *output injection* respectively.

2.1 Invariants and Restricted Maps

Definition 2.1. (Invariant). Given a linear map $A : \mathcal{X} \to \mathcal{X}$, a subspace $\mathcal{L} \subseteq \mathcal{X}$ is an *A-invariant* (or, simply, an *invariant* when the map referred to is clear from the context) if

$$A\mathcal{L} \subseteq \mathcal{L} \qquad (2.3)$$

Let L be a basis matrix of $\mathcal{L}$: the following statements are equivalent to (2.3):

- a matrix X exists such that $A L = L X$; $\qquad (2.4)$
- $\mathcal{L}$ is a locus of trajectories of the free system $\dot{x}(t) = A x(t)$. $\qquad (2.5)$

Given the subspaces $\mathcal{D}$, $\mathcal{E}$ contained in $\mathcal{X}$ and such that $\mathcal{D} \subseteq \mathcal{E}$, and a linear map $A : \mathcal{X} \to \mathcal{X}$, it is well known that the set of all A-invariants $\mathcal{L}$ satisfying $\mathcal{D} \subseteq \mathcal{L} \subseteq \mathcal{E}$ is a nondistributive lattice with respect to $\subseteq, +, \cap$. We denote with $\max\mathcal{L}(A, \mathcal{E})$ the maximal A-invariant contained in $\mathcal{E}$ (the sum of all the A-invariants contained in $\mathcal{E}$) and with $\min\mathcal{L}(A, \mathcal{D})$ the minimal A-invariant containing $\mathcal{D}$ (the intersection of all the A-invariants containing $\mathcal{D}$): the above lattice is non-empty if and only if $\mathcal{D} \subseteq \max\mathcal{L}(A, \mathcal{E})$ or $\min\mathcal{L}(A, \mathcal{D}) \subseteq \mathcal{E}$.

The restriction of map A to the A-invariant subspace $\mathcal{L}$ is denoted by $A|_\mathcal{L}$; $\mathcal{L}$ is said to be *internally stable* if $A|_\mathcal{L}$ is stable. Given two A-invariants $\mathcal{L}_1$ and $\mathcal{L}_2$ such that $\mathcal{L}_1 \subseteq \mathcal{L}_2$, the map induced by A on the quotient space $\mathcal{L}_2/\mathcal{L}_1$ is denoted by $A|_{\mathcal{L}_2/\mathcal{L}_1}$. In particular, an A-invariant $\mathcal{L}$ is said to be *externally stable* if $A|_{\mathcal{X}/\mathcal{L}}$ is stable.

Definition 2.2. (Complementable Invariant). An A-invariant $\mathcal{L} \subseteq$

$\mathcal{X}$ is said to be *complementable* if an A-invariant $\mathcal{L}_c$ exists such that $\mathcal{L} \oplus \mathcal{L}_c = \mathcal{X}$; if so, $\mathcal{L}_c$ is called a *complement* of $\mathcal{L}$.

The importance of invariants in connection with the triple (A, B, C) is related to the following well known, basic properties:

Property 2.1. The reachability subspace of the pair (A, B), i.e., the set of all the states that can be reached from the origin in any finite time by means of admissible (piecewise continuous) control actions, is $\mathcal{R} = \min\mathcal{L}(A, B)$. If $\mathcal{R} = \mathcal{X}$, (A, B) is said to be *reachable*.

Property 2.2. The pair (A, B) is pole-assignable with state feedback if and only if $\mathcal{R} = \mathcal{X}$, stabilizable with state feedback if and only if $\mathcal{R}$ is externally stable.

Property 2.3. The unobservability subspace of the pair (C, A), i.e., the set of all the initial states that cannot be recognized from the output function, is $\mathcal{Q} = \max\mathcal{L}(A, C)$. If $\mathcal{Q} = \{0\}$, (C, A) is said to be *observable*.

Property 2.4. The pair (C, A) is pole-assignable with output injection if and only if $\mathcal{Q} = \{0\}$, stabilizable with output injection if and only if $\mathcal{Q}$ is internally stable.

Algorithm 2.1. (Minimal A-invariant containing $\mathcal{B}$). The subspace $\min\mathcal{L}(A, \mathcal{B})$ coincides with the last term of the sequence

$$\begin{aligned} \mathcal{Z}_0 &:= \mathcal{B} \\ \mathcal{Z}_i &:= \mathcal{B} + A\,\mathcal{Z}_{i-1} \qquad (i = 1, \dots, k) \end{aligned} \qquad (2.6)$$

where the value of $k \leq n-1$ is determined by condition $\mathcal{Z}_{k+1} = \mathcal{Z}_k$.

Algorithm 2.2. (Maximal A-invariant contained in $\mathcal{C}$). The subspace $\max\mathcal{L}(A, \mathcal{C})$ satisfies

$$\max\mathcal{L}(A, \mathcal{C}) = \min\mathcal{L}(A^T, \mathcal{C}^\perp)^\perp \qquad (2.7)$$

hence it can be computed by using Algorithm 2.1 again.

Algorithm 2.3. (Internal and External Eigenstructure of an A-invariant). Matrices P and Q representing $A|_{\mathcal{L}}$ and $A|_{\mathcal{X}/\mathcal{L}}$ up to an isomorphism, are derived as follows. Let us consider the similarity transformation $T := [L \ T_2]$, with $\text{im}\,L = \mathcal{L}$ (L is a basis matrix of $\mathcal{L}$) and T_2 such that T is nonsingular. In the new basis the linear transformation A is expressed by

$$A' = T^{-1}A\,T = \begin{bmatrix} A'_{11} & A'_{12} \\ O & A'_{22} \end{bmatrix} \qquad (2.8)$$

The requested matrices are defined as $P := A'_{11}$, $Q := A'_{22}$.

Algorithm 2.4. (Complementability of an A-invariant). Let us consider again the change of basis introduced in Algorithm 2.3. $\mathcal{L}$ is complementable if and only if the *Sylvester equation*[1] in X

$$A'_{11} X - X A'_{22} = -A'_{12} \qquad (2.9)$$

admits a solution. If so, a basis matrix of $\mathcal{L}_c$ is given by $L_c := L X + T_2$.[2]

The following property states a sufficient condition for complementability that is often applied in regulation theory. In fact, if this condition is satisfied, equation (2.9) admits a unique solution.

Property 2.5. An A-invariant $\mathcal{L}$ is complementable and admits a unique complement $\mathcal{L}_c$ if the linear mappings $A|_{\mathcal{L}}$ and $A|_{\mathcal{X}/\mathcal{L}}$ (hence matrices P and Q provided by Algorithm 2.3) have no common eigenvalues, i.e., if

$$\sigma(A|_{\mathcal{L}}) \cap \sigma(A|_{\mathcal{X}/\mathcal{L}}) = \emptyset \qquad (2.10)$$

2.2 Controlled Invariants and Self-Bounded Controlled Invariants

Definition 2.3. (Controlled Invariant). Given a linear map $A : \mathcal{X} \to \mathcal{X}$ and a subspace $\mathcal{B} \subseteq \mathcal{X}$ a subspace $\mathcal{V} \subseteq \mathcal{X}$ is an $(A, \mathcal{B})$-*controlled invariant* (or, simply, a *controlled invariant* if A and $\mathcal{B}$ are clear from the context) if

$$A \mathcal{V} \subseteq \mathcal{V} + \mathcal{B} \qquad (2.11)$$

Let B and V be basis matrices of $\mathcal{B}$ and $\mathcal{V}$ respectively: the following statements are equivalent to (2.11):

- a matrix F exists such that $(A+BF) \mathcal{V} \subseteq \mathcal{V}$; (2.12)
- matrices X and U exist such that $A V = V X + B U$; (2.13)

---

[1]The generic Sylvester equation $A X + X B = C$, where A is $m \times m$, B is $n \times n$ and X, C are both $m \times n$, is equivalent to a set of mn linear equation. It has a unique solution if $\sigma(A) \cap \sigma(B) = \emptyset$.

[2]A sketch of the proof. Let m be the dimension of $\mathcal{L}$. In the new coordinate system the matrices

$$L' = \begin{bmatrix} I_m \\ O \end{bmatrix} \quad \text{and} \quad L'_c = \begin{bmatrix} X \\ I_{n-m} \end{bmatrix}$$

are basis matrices of $\mathcal{L}$ and a complement $\mathcal{L}_c$ respectively, since $[L' \ L'_c]$ is clearly nonsingular and both satisfy (2.4). The corresponding basis matrices in the original basis are $T L'$ and $T L'_c$.

- $\mathcal{V}$ is a locus of trajectories of the pair (A, B).   (2.14)

The last statement is very important in connection with control problems: a subspace $\mathcal{V} \subseteq \mathcal{X}$ is an (A, B)-controlled invariant if and only if, starting from any initial state belonging to $\mathcal{V}$, it is possible to follow a state trajectory completely belonging to $\mathcal{V}$.

It is easily shown that the sum of any two controlled invariants is a controlled invariant; thus the set of all controlled invariants contained in a given subspace $\mathcal{E} \subseteq \mathcal{X}$ is a semilattice with respect to $\subseteq, +$, hence admits a supremum, the maximal (A, B)-controlled invariant contained in $\mathcal{E}$, that is denoted by $\max \mathcal{V}(A, B, \mathcal{E})$ (or simply $\mathcal{V}^*$ if the matrices and subspaces involved in the definition are clear from the context).

Referring to the pair (A, B), we denote with $\mathcal{R}_\mathcal{V}$ the reachable subspace from the origin by trajectories constrained to belong to $\mathcal{V}$. By virtue of (2.12) it is derived as $\mathcal{R}_\mathcal{V} = \min \mathcal{L}(A + BF, \mathcal{V} \cap B)$ and, being clearly an $(A + BF)$-invariant, it also is an (A, B)-controlled invariant.

A generic (A, B)-controlled invariant $\mathcal{V}$ is said to be *internally stabilizable* or *externally stabilizable* if at least one matrix F exists such that $(A + BF)|_\mathcal{V}$ is stable or at least one matrix F exists such that $(A + BF)|_{\mathcal{X}/\mathcal{V}}$ is stable. It is easily proved that the eigenstructure of $(A + BF)|_{\mathcal{V}/\mathcal{R}_\mathcal{V}}$ is independent of F; it is called the *internal unassignable eigenstructure of $\mathcal{V}$*: $\mathcal{V}$ is both internally and externally stabilizable with the same F if and only if its internal unassignable eigenstructure is stable and the A-invariant $\mathcal{V} + \min \mathcal{L}(A, B)$ is externally stable. Hence, referring to the pair (A, B), external stabilizability of any controlled invariant is guaranteed if (A, B) is stabilizable: under this assumption we can state that any (A, B)-controlled invariant is stabilizable if and only if its internal unassignable eigenstructure is stable.

Definition 2.4. (Self-Bounded Controlled Invariant). Given a linear map $A : \mathcal{X} \to \mathcal{X}$ and two subspaces $B \subseteq \mathcal{X}$, $\mathcal{E} \subseteq \mathcal{X}$, a subspace $\mathcal{V} \subseteq \mathcal{X}$ is an (A, B)-controlled invariant *self-bounded with respect to $\mathcal{E}$* if, besides (2.11), the following relations hold

$$\mathcal{V} \subseteq \mathcal{V}^* \subseteq \mathcal{E} \tag{2.15}$$

$$\mathcal{V}^* \cap B \subseteq \mathcal{V} \tag{2.16}$$

The set of all (A, B)-controlled invariants self-bounded with respect to $\mathcal{E}$ is a nondistributive lattice with respect to $\subseteq, +, \cap$, whose supremum is $\mathcal{V}^*$ and whose infimum is $\mathcal{R}_{\mathcal{V}^*}$.

Given subspaces $\mathcal{D}$, $\mathcal{E}$ contained in $\mathcal{X}$ and such that $\mathcal{D} \subseteq \mathcal{V}^*$, the infimum of the lattice of all (A, B)-controlled invariants self-bounded with respect to $\mathcal{E}$ and containing $\mathcal{D}$ is the reachable set restricted to $\mathcal{V}^*$ with forcing action $\mathcal{B}+\mathcal{D}$, i.e., $\min \mathcal{L}(A+BF, \mathcal{V}^* \cap \mathcal{B}+\mathcal{D})$, with F such that $(A+BF)\mathcal{V}^* \subseteq \mathcal{V}^*$. The following property makes the concept of self-boundedness very interesting in connection with synthesis procedures with stability requirements.

Property 2.6. Let $\mathcal{D} \subseteq \mathcal{V}^* = \max\mathcal{V}(A,\mathcal{B},\mathcal{E})$. If the infimum of the lattice of all $(A,\mathcal{B})$-controlled invariants self-bounded with respect to $\mathcal{E}$ and containing $\mathcal{D}$ is not internally stabilizable, no other $(A,\mathcal{B})$-controlled invariant contained in $\mathcal{E}$ and containing $\mathcal{D}$ is internally stabilizable.

2.3 Conditioned Invariants

Definition 2.5. (Conditioned Invariant). Given a linear map $A : \mathcal{X} \to \mathcal{X}$ and a subspace $\mathcal{C} \subseteq \mathcal{X}$ a subspace $\mathcal{S} \subseteq \mathcal{X}$ is an $(A,\mathcal{C})$-*conditioned invariant* (or, simply, a *conditioned invariant* when A and $\mathcal{C}$ are clear from the context) if

$$A(\mathcal{S} \cap \mathcal{C}) \subseteq \mathcal{S} \tag{2.17}$$

The following statement is equivalent to (2.17):

- a matrix G exists such that $(A+GC)\mathcal{S} \subseteq \mathcal{S}$. $\tag{2.18}$

The intersection of any two conditioned invariants is a conditioned invariant; thus the set of all conditioned invariants containing a given subspace $\mathcal{D} \subseteq \mathcal{X}$ is a semilattice with respect to $\subseteq, \cap$, hence admits an infimum, the minimal $(A,\mathcal{C})$-conditioned invariant containing $\mathcal{D}$, that is denoted by $\min\mathcal{S}(A,\mathcal{C},\mathcal{D})$ (or simply $\mathcal{S}^*$ if the matrices and subspaces involved in the definition are clear from the context).

Controlled and conditioned invariants are dual. Controlled invariants are used in control problems, while conditioned invariants are used in observation problems. The orthogonal complement of an $(A,\mathcal{C})$-conditioned invariant is an $(A^T,\mathcal{C}^\perp)$-controlled invariant, hence the orthogonal complement of an $(A,\mathcal{C})$-conditioned invariant containing a given subspace $\mathcal{D}$ is an $(A^T,\mathcal{C}^\perp)$-controlled invariant contained in $\mathcal{D}^\perp$. Furthermore, the infimum of the lattice of all $(A,\mathcal{B})$-controlled invariants self-bounded with respect to a given subspace $\mathcal{E}$ can be expressed in terms of conditioned invariants as follows.

Property 2.7. Let $\mathcal{D} \subseteq \mathcal{V}^* = \max\mathcal{V}(A,\mathcal{B},\mathcal{E})$. The infimum of the lattice of all $(A,\mathcal{B})$-controlled invariants self-bounded with respect to $\mathcal{E}$ and containing $\mathcal{D}$ is expressed by

$$\max\mathcal{V}(A,\mathcal{B},\mathcal{E}) \cap \min\mathcal{S}(A,\mathcal{E},\mathcal{B}+\mathcal{D}) \tag{2.19}$$

Note, in particular, that $\mathcal{R}_{\mathcal{V}^*} = \mathcal{V}^* \cap \mathcal{S}^*$, with $\mathcal{S}^* := \min \mathcal{S}(A, \mathcal{E}, B)$.

Algorithm 2.5. (Minimal (A, C)-conditioned invariant containing B). The subspace $S^* = \min \mathcal{S}(A, C, B)$ coincides with the last term of the sequence

$$
\begin{aligned}
\mathcal{Z}_0 &:= B \\
\mathcal{Z}_i &:= B + A\left(\mathcal{Z}_{i-1} \cap C\right) \qquad (i = 1, \ldots, k)
\end{aligned}
\tag{2.20}
$$

where the value of $k \le n-1$ is determined by condition $\mathcal{Z}_{k+1} = \mathcal{Z}_k$.

Algorithm 2.6. (Maximal (A, B)-controlled invariant contained in $\mathcal{E}$). The subspace $\mathcal{V}^* = \max \mathcal{V}(A, B, \mathcal{E})$ satisfies

$$
\max \mathcal{V}(A, B, \mathcal{E}) = \min \mathcal{S}(A^T, B^\perp, \mathcal{E}^\perp)^\perp
\tag{2.21}
$$

hence it can be computed by using Algorithm 2.5.

Algorithm 2.7. (Matrix F such that $(A+BF)\mathcal{V} \subseteq \mathcal{V}$). Let V be a basis matrix of the (A, B)-controlled invariant $\mathcal{V}$. Matrices X and U satisfying (2.13) are derived with

$$
\begin{bmatrix} X \\ U \end{bmatrix} = [V \, B]^+ \, A V
\tag{2.22}
$$

where the symbol $^+$ denotes the pseudoinverse. Then, assume

$$
F := -U \, (V^T V)^{-1} V^T
\tag{2.23}
$$

Matrix F satisfies $(A+BF)V = VX$, hence, by the equivalence of (2.3) and (2.4), $(A+BF)\mathcal{V} \subseteq \mathcal{V}$.

Algorithm 2.8. (Internal unassignable eigenstructure of an (A, B)-controlled invariant). A matrix P representing the map $(A+BF)|_{\mathcal{V}/\mathcal{R}_{\mathcal{V}}}$ up to an isomorphism, is derived as follows. Let us consider the similarity transformation $T := [T_1 \, T_2 \, T_3]$, with $\mathrm{im} \, T_1 = \mathcal{R}_{\mathcal{V}}$, $\mathrm{im} \, T_2 = \mathcal{V}$ and T_3 such that T is nonsingular. In the new basis matrix $A+BF$ is expressed by

$$
(A+BF)' = T^{-1}(A+BF)\, T = \begin{bmatrix} A'_{11} & A'_{12} & A'_{13} \\ O & A'_{22} & A'_{23} \\ O & O & A'_{33} \end{bmatrix}
\tag{2.24}
$$

The requested matrix is $P := A'_{22}$.

2.4 System Invertibility and Perfect Output Controllability

The concepts of system invertibility and output controllability of the triple (A, B, C) are of paramount importance when approaching problems related to perfect tracking.

Definition 2.6. (Invertibility of a triple). Let us refer to the triple (A, B, C), where B is assumed to have maximum rank, and consider the linear operator $T_f : \mathcal{U}_f \to \mathcal{Y}_f$, from the vector space of the admissible input functions to the vector space of the zero-state responses, expressed by the convolution integral

$$y(t) = C \int_0^t e^{A(t-\tau)} B\, u(\tau)\, d\tau \tag{2.25}$$

(A, B, C) is said to be *invertible* (or *zero-state invertible)*) if $\ker T_f = \{0\}$.

Let $\mathcal{V}^* := \max \mathcal{V}(A, \mathcal{B}, \mathcal{C})$ with $\mathcal{B} := \operatorname{im} B$, $\mathcal{C} := \ker C$, and denote, as before, by $\mathcal{R}_{\mathcal{V}^*}$ the reachable subspace on $\mathcal{V}^*$: clearly $\ker T_f = \{0\}$ if and only if $\mathcal{R}_{\mathcal{V}^*} = \{0\}$. Hence the following property holds.

Property 2.8. The triple (A, B, C), where B is assumed to be maximum-rank, is invertible (zero-state invertible).[3] if

$$\mathcal{V}^* \cap \mathcal{B} = \{0\} \tag{2.26}$$

Definition 2.7. (Functional Controllability). Refer to the operator (2.25) and denote by $\mathcal{Y}_f^{(n)}$ the set of all "smooth enough" output functions, i.e., of all output functions whose derivative of order n or more is piecewise continuous. (A, B, C) is said to be *functionally controllable* if $\mathcal{Y}_f^{(n)} \subseteq \operatorname{im} T_f$.[4]

The following property is easily derived from Property 2.8 by a simple duality argument.

---

[3]A "strong" system invertibility refers to the linear operator $T_f' : \mathcal{X} \times \mathcal{U}_f \to \mathcal{Y}_f$ defined by

$$y(t) = C e^{At} x_0 + C \int_0^t e^{A(t-\tau)} B\, u(\tau)\, d\tau$$

whose null space is clearly zero if and only if $\mathcal{V}^* = \{0\}$.

[4]It is worth noting that in the system theory literature zero-state invertible systems are often called *left-invertible* ("left" since based on the kernel of the input-to-output functional map being zero) and zero-state functionally controllable systems are often called *right-invertible* ("right" since based on the image of the same functional map being the full space).

Property 2.9. Let $S^* := \min S(A, C, B)$ with $B := \mathrm{im}B$, $C := \ker C$. The triple (A, B, C), where C is assumed to be maximum-rank, is functionally controllable if

$$S^* + C = X \tag{2.27}$$

In order to state a very neat extension to MIMO systems of the concept of relative degree that characterizes the output function smoothness in the SISO case, let us introduce the following extension of functional output controllability, that refers to a subspace of the output space and to the generic h-th derivative of the output function instead of the n-th derivative.

Definition 2.8. (Constrained Functional Controllability). A subspace $\mathcal{Y}^{(h)} \subseteq \mathbf{R}^q$ is said to be a *functional output controllability subspace with respect to the h-th derivative* if the output of the triple (A, B, C) can be driven along any trajectory $y(\cdot)$ such that $y(t) \in \mathcal{Y}^{(h)}$ for all t and with the h-th derivative piecewise continuous.

This is possible if and only if at least one (A, B)-controlled invariant $\mathcal{V} \subseteq X$ exists such that $\mathcal{Y}^{(h)} = C\mathcal{V}$ and that, for every initial state $x_0 \in \mathcal{V}$, it is possible to drive the state along a trajectory on $\mathcal{V}$ that maps into the given $y(\cdot)$ in the output space. The following characterizing property holds.

Property 2.10. Let us refer to (A, B, C). An (A, B)-controlled invariant $\mathcal{V} \subseteq X$ corresponds to a functional output controllability subspace with respect to the h-th derivative $\mathcal{Y}^{(h)}$ if

$$\mathcal{Y}^{(h)} = C\mathcal{V} \tag{2.28}$$
$$\mathcal{V} \subseteq \mathcal{V} \cap \mathcal{Z}_{h-1} + C \tag{2.29}$$

with $\mathcal{Z}_{h-1}$ defined by

$$\begin{aligned}
\mathcal{Z}_0 &:= B \\
\mathcal{Z}_j &:= B + A\,(\mathcal{Z}_{j-1} \cap \mathcal{V} \cap C) \qquad (j = 1, \ldots, h-1)
\end{aligned} \tag{2.30}$$

where, as before, $C := \ker C$ and $B := \mathrm{im}B$.

Property 2.10 is consistent with Property 2.9: in fact, the whole output space $\mathbf{R}^q$ satisfies Property 2.10 with $h = n$: in this case the stated conditions are satisfied with $\mathcal{V} = X$, hence sequence (2.30) provides S^*, so that (2.29) is equivalent to (2.27).

Algorithm 2.9. (Maximum Subspace of Contrained Functional

Output Controllability). Let us refer to (A, B, C). Given a subspace $\mathcal{Y}^{(h)} \subseteq \mathbf{R}^q$, let $\mathcal{E} := C^{-1} \mathcal{Y}^{(h)}$; the maximal (A, B)-controlled invariant $\mathcal{V}_{\mathcal{E}}^{(h)} \subseteq \mathcal{X}$ contained in $\mathcal{E}$ such that the output can be driven on $C \mathcal{V}_{\mathcal{E}}^{(h)} \subseteq \mathcal{Y}^{(h)}$ along any trajectory $y(t)$ with piecewise continuous h-th derivative for all the initial states $x(0) \in \mathcal{V}_{\mathcal{E}}^{(h)}$, is the last term (with identity of two consecutive terms as the stop condition) of the sequence

$$\begin{aligned} \mathcal{V}_0 &= \mathcal{E} \\ \mathcal{V}_i &= \max \mathcal{V}(A, B, \mathcal{V}_{i-1} \cap \mathcal{Z}_{i,h-1} + C) \qquad (i = 1, 2, \ldots) \end{aligned} \qquad (2.31)$$

with $\mathcal{Z}_{i,h-1}$ defined by the recursion process

$$\begin{aligned} \mathcal{Z}_{i,0} &= \mathcal{B} \\ \mathcal{Z}_{i,j} &= \mathcal{B} + A(\mathcal{Z}_{i,j-1} \cap \mathcal{V}_{i-1} \cap C) \qquad (j = 1, \ldots, h-1) \end{aligned} \qquad (2.32)$$

where, as before, $C := \ker C$ and $\mathcal{B} := \operatorname{im} B$.

Algorithm 2.10. (Multivariable Relative Degree). The relative degree ρ_i of output y_i, $(i = 1, \ldots, q)$ is obtained with Algorithm 2.9 by assuming $\mathcal{Y}^{(h)} := \{y : y_k = 0, k = 1, \ldots, q, k \neq i\}$ (the i-th coordinate axis) and repeating for $h = 1, 2, \ldots$ until equality $\mathcal{Y}^{(h)} = C \mathcal{V}_{\mathcal{E}}^{(h)}$ is obtained. When this occurs, we have $\rho_i = h$.

2.5 Invariant Zeros

Invariant zeros are characteristic parameters of the triple (A, B, C) that significantly affect solvability of both the standard asymptotic robust regulation problem, presented in Section 3, and the perfect tracking problem presented in Section 4. Roughly speaking, an invariant zero corresponds to a mode that, if suitably injected at the input of a dynamic system, can be nulled at the output by a suitable choice of the initial state.[5]

Definition 2.9. (Invariant Zero and Invariant Zero Structure). The *invariant zeros* of (A, B, C) are the internal unassignable eigenvalues of

---

[5]This is particularly clear in the SISO case. In fact, the second term on the left of

$$\frac{1}{s - z_i} \frac{K(s - z_1) \ldots (s - z_m)}{(s - p_1) \ldots (s - p_n)} - \frac{K(s - z_1) \ldots (s - z_{i-1})(s - z_{i+1}) \ldots (s - z_m)}{(s - p_1) \ldots (s - p_n)} = 0$$

being a strictly proper rational function with all poles equal to those of the system, can be interpreted as a free motion. It is shown here by using a geometric-type argument that a similar property holds in the MIMO case, but the mode corresponding to any invariant zero must be distributed, with suitable coefficients, on all the inputs.

$\mathcal{V}^* := \max\mathcal{V}(A, B, C)$. The *invariant zero structure* of (A, B, C) is the internal unassignable eigenstructure of $\mathcal{V}^*$.

Thus, the invariant zeros are the eigenvalues of matrix P derived with Algorithm 2.8, while the invariant zeros structure is the eigenstructure of matrix P itself.

Property 2.11. Let W be a real $m \times m$ matrix having the invariant zero structure of (A, B, C) as eigenstructure. There exist a real $p \times m$ matrix L and a real $n \times m$ matrix X, with (X, W) observable, such that applying to (A, B, C) the input function

$$u(t) = L\, e^{Wt}\, v_0 \tag{2.33}$$

where $v_0 \in \mathbf{R}^m$ denotes an arbitrary column vector, and starting from the initial state $x_0 = X v_0$, the output $y(\cdot)$ is identically zero, while the state evolution (on $\ker C$) is described by

$$x(t) = X\, e^{Wt}\, v_0 \tag{2.34}$$

The observability of (X, W) implies that all the modes corresponding to the zero structure can be excited with a suitable choice of v_0; since (C, A) is observable and the output is identically zero all these modes are necessarily injected at the input through L.

Note that the application of the mode corresponding to a single zero is included in the statement; since v_0 is arbitrary, W is defined within an isomorphism, hence it can be assumed to be in the real Jordan form without any loss of generality. Thus, the modes corresponding to its eigenvalues, both simple and multiple, can be easily individually excited by suitably setting to nonzero values only the components of v_0 that represent the initial conditions of these modes.

In plain words Property 2.11 means that any linear combination of the modes corresponding to the invariant zero structure, if suitably injected at the input, can be nulled at the output by a suitable choice of the initial state.

By substituting (2.33) and (2.34) in (2.1) and (2.2) the following equations for L and X are immediately obtained:

$$AX - XW \;=\; -BL \tag{2.35}$$
$$CX \;=\; O \tag{2.36}$$

Algorithm 2.11. (Matrices L and X defined in Property 2.11). First compute a state feedback matrix F such that $(A+BF)\mathcal{V}^* \subseteq \mathcal{V}^*$. F

also allows arbitrary assignment of all the eigenvalues except the internal unassignable eigenvalues of $\mathcal{V}^*$, so we can assume that no other eigenvalue of $A+BF$ is equal to them. On this assumption $\mathcal{R}_{\mathcal{V}^\bullet}$ is an $(A+BF)$-invariant complementable with respect to $\mathcal{V}^*$: hence an (A,B)-controlled invariant $\mathcal{V}$ exists such that $\mathcal{R}_{\mathcal{V}^\bullet} \oplus \mathcal{V} = \mathcal{V}^*$, $(A+BF)\mathcal{V} \subseteq \mathcal{V}$. Consider the change of basis defined by the transformation $T := [T_1\, T_2\, T_3]$, with $\mathrm{im}\, T_1 = \mathcal{R}_{\mathcal{V}^\bullet}$, $\mathrm{im}\, T_2 = \mathcal{V}$. With respect to the new basis we obtain the structure

$$T^{-1}(A+BF)\,T = \begin{bmatrix} A'_{11} & O & A'_{13} \\ O & A'_{22} & A'_{23} \\ O & O & A'_{33} \end{bmatrix} \tag{2.37}$$

Note that the invariant zeros are the eigenvalues of A'_{22} and the invariant zero structure is the eigenstructure of A'_{22}. Then, assume $L := FT_2$, $X := T_2$.[6]

2.6 A classical application: the Disturbance Localization Problem

The disturbance localization problem is one of the earliest (1969) applications of controlled invariants, and since it played the role of the reference problem for introducing new tools (like, for instance, self-bounded controlled invariants) it is here briefly recalled. Let us consider the system

$$\dot{x}(t) = A\,x(t) + B\,u(t) + D\,d(t) \tag{2.38}$$
$$y(t) = C\,x(t) \tag{2.39}$$

where u denotes the manipulable input, d the disturbance input, that may be either *inaccessible* of *accessible* for measurement.

Problem 2.1. (Disturbance Localization). Referring to system (2.38, 2.39), determine, if possible, a state-to-input feedback matrix F, and, if disturbance is accessible, a disturbace-to-input matrix S, such that, starting at the zero state, any disturbance function $d(\cdot)$ does not affect the output y.

---

[6]A sketch of the proof. Since W is defined within an isomorphism, it is possible to assume $W = A'_{22}$ without loss of generality, so that (2.35) becomes $A\,T_2 - T_2\,A'_{22} = -B\,F\,T_2$, or $(A+BF)\,T_2 = T_2\,A'_{22}$, that directly follows from (2.37). Since $\mathrm{im}\,X = \mathcal{V} \subseteq \mathcal{V}^*$, (2.36) is satisfied. Furthermore, the pair (X, W) is observable since the rank of X is maximal and equal to the dimension of W.

The system with state feedback is described by

$$\dot{x}(t) = (A + BF)\, x(t) + D_1\, d(t) \tag{2.40}$$
$$y(t) = C\, x(t) \tag{2.41}$$

with $D_1 := D$ if disturbance is inaccessible or $D_1 := D + BS$, where S is a suitable matrix to be determined, if disturbance is accessible. It behaves as requested if and only if its reachable set by d, i.e., the minimum $(A + BF)$-invariant containing $\mathcal{D}_1 := \operatorname{im} D_1$, is contained in $\mathcal{C} := \ker C$.

Let us define $\mathcal{V}^* := \max \mathcal{V}(A, B, C)$. Since any $(A + BF)$-invariant is an (A, B)-controlled invariant, the inaccessible disturbance localization problem has a solution if and only if the following *structural condition* holds:

$$\mathcal{D} \subseteq \mathcal{V}^* \tag{2.42}$$

while in the case of the accessible disturbance localization problem, inclusion (2.42) is replaced by

$$\mathcal{D} \subseteq \mathcal{V}^* + \mathcal{B} \tag{2.43}$$

Conditions (2.42) and (2.43) are constructive in the sense that, if it they are satisfied, Algorithm 2.7 directly provides a matrix F that solves the problem of inaccessible disturbance localization, while matrix S is such that $D + BS$ is the projection of d on $\mathcal{V}^*$ along $\mathcal{B}$. However, it is technically sound to require stability besides disturbance localization. The disturbance localization problem with stability is stated as follows.

Problem 2.2. (Disturbance Localization with Stability). Referring to system (2.38, 2.39), where (A, B) is assumed to be stabilizable, determine, if possible, a state-to-input feedback matrix F, and, if disturbance is accessible, a disturbace-to-input matrix S, such that, starting at the zero state, any disturbance function $d(\cdot)$ does not affect the output y and the system matrix $A + BF$ is strictly stable.

Let us suppose that (2.42) and (2.43), clearly still necessary, are statisfied and denote by

$$\mathcal{V}_m := \mathcal{V}^* \cap \mathcal{S}^* = \max \mathcal{V}(A, B, C) \cap \min \mathcal{S}(A, C, B + D) \tag{2.44}$$

the minimum controlled invariant self-bounded with respect to $\mathcal{C}$ and satisfying (2.42) or (2.43), that can be also computed by using the relation

$$\mathcal{V}_m := \min \mathcal{L}(A + BF, \mathcal{V}^* \cap (\mathcal{B} + \mathcal{D})) \text{ with } F \text{ such that } (A + BF)\, \mathcal{V}^* \subseteq \mathcal{V}^* \tag{2.45}$$

From Property 2.6 the following statement is directly obtained.

Property 2.12. The disturbance localization problem with stability is solvable if and only if (2.43) or (2.43) holds and V_m is internally stabilizable.

Note that solvability of the problem is expressed by two types of conditions: the first is a *structural condition* and the second a *stabilizability condition*. This feature is also exhibited by the other synthesis problems approached in the next sections of this monography. The stabilizability condition can be expressed in terms of invariant zeros as follows.

Property 2.13. Let $\mathcal{Z}(u; y)$ be the set of all the invariant zeros of (38, 39) referred to input u and $\mathcal{Z}(u, d; y)$ that referred to inputs u, d, considered as a whole. The disturbance localization problem with stability is solvable if and only if (2.42) or (2.43) holds and the elements of $\mathcal{Z}(u; y)$ not belonging also to $\mathcal{Z}(u, d; y)$ are all stable.

Property 2.13 is derived from Property 2.12 as follows: let $\mathcal{R}_{V^*}$ be the reachable set on V^* by u, while V_m coincides with the reachable set on V^* by u and d, both considered as manipulable inputs; both are clearly $(A+BF)$-invariants. The internal unassignable eigenvalues of V_m are the elements of $\sigma(A+BF|_{V_m/\mathcal{R}_{V^*}})$, while the set of zeros considered in the statement are defined by $\mathcal{Z}(u; y) := \sigma(A+BF|_{V^*/\mathcal{R}_{V^*}})$ and $\mathcal{Z}(u, d; y) := \sigma(A+BF)|_{V^*/\mathcal{R}_{V_m}})$.

The main criticism to which the geometric techniques are subject is that they generally provide non-robust solutions. The disturbance localization problem is emblematic with respect to this: small variations of the controlled system parameters may destroy the structural feature consisting of a very vulnerable subspace inclusion.

In the next section it is shown that robust multivariable disturbance rejection can be achieved by feedback and studied again with geometric techniques, but within relaxed specifications: disturbance is modelled by an exosystem (reproducing, for instance, a step) and asymptotic robust rejection is achieved with a suitable feedback regulator. Since the error dynamics is arbitrary under standard controllability and observability assumptions, the effect of disturbance on regulated output may be made arbitrarily small in any finite bandwidth, but perfect rejection is obtained only at the steady state.

2.7   Computational support with Matlab

The algebraic operations concerning subspaces required to implement the algorithms of the geometric approach considered in this section are easily performed on the corresponding basis matrices by using some Matlab routines available in a diskette included with Basile and Marro (1992). Those directly related to the previously presented concepts are:

Q = ima(A,p) Orthonormalization.

Q = ortc(A) Complementary orthogonalization.

Q = sums(A,B) Sum of subspaces.

Q = ints(A,B) Intersection of subspaces.

Q = invt(A,X) Inverse transform of a subspace.

Q = ker(A) Kernel of a matrix.

Q = mininv(A,B) Minimal A-invariant containing im$B$.

Q = maxinv(A,X) Maximal A-invariant contained in im$X$.

[P,Q] = stabi(A,X) Matrices for the internal and external stability of the A-invariant im$X$.

Q = mainco(A,B,X) Maximal (A, B)-controlled invariant contained in im$X$.

Q = miinco(A,C,X) Minimal (A, C)-conditioned invariant containing im$X$.

F = effe(A,B,X) State feedback matrix such that the (A, B)-controlled invariant im$X$ is an $(A+BF)$-invariant.

[P,Q] = stabv(A,B,X) Matrices for the internal and external stabilizability of the (A, B)-controlled invariant im$X$.

z = gazero(A,B,C,D) Invariant zeros of (A, B, C) or (A, B, C, D).

Let us complete the above computational machinery with the following m-file, for constrained functional output controllability, hence useful to extend the concept of relative degree to the multivariable case.

```
function W = maxpcs(A,B,C,E,ii)
%    W = maxpcs(A,B,C,E,i): maximal perfect
%    controllability subspace with respect to
%    the i-th derivative contained in ker(E).

%    G.Marro \& A.Piazzi 4-30-95
```

```
nv=length(A);
W=ker(E);
[mw,nw]=size(W);
B1=ima(B,0);
C1=ker(C);
h=0;
while (nv-nw) > 0 | h==0
  nv=nw;
  Z=B1;
  for jj=1:(ii-1)
    Z=ima([B1,A*ints(ints(Z,W),C1)],0);
  end
  W=mainco(A,B,sums(ints(W,Z),C1));
  [mw,nw]=size(W);
end
% --- last line of maxpcs ---
```

Notes and References. Most of the previously presented concepts and algorithms are reported in the books by Wonham [55] and Basile and Marro [11]. The geometric characterization of system invertibility and functional controllability is due to Morse and Wonham [44], while constrained output controllability, hence the geometric-type extension of the relative degree concept to multivariable system, was introduced by Basile and Marro [8]. The elegant algorithm providing the reachable subspace on a controlled invariant by intersection with a conditioned invariant is due to Morse [43], and the role of invariant zeros in regulation problems was pointed out by Francis and Wonham [27]. The disturbance localization problem both with state and output feedback was presented by Basile and Marro as the first application of the geometric techniques in system synthesis [6], while the complete solution of this problem with output dynamic feedback and stability was given by Willems and Commault [53], and disturbance localization plus regulation by Schumacher [47]. Self-bounded controlled invariants were introduced by Basile, Marro and Piazzi [9, 12, 13] to deal with stabilizability in a purely geometric context, and connection between invariant zeros and stabilizability in some well-known synthesis problems, including disturbance localization, was investigated by Marro and Piazzi [38].

3   Steady-State Robust Regulation with Feedback and the Internal Model

This section extends some well-known properties of SISO feedback regulators to the MIMO case by using the previously introduced geometric techniques. The features of regulators considered here are: *(i)* closed-loop asymptotic stability or, more generally, pole assignability and *(ii)* the use of an internal model to achieve asymptotic tracking of the reference input and rejection of the disturbance input under the assumption that both reference and disturbance are generated by linear exosystems. Since the overall system considered, included the exosystems, is described by a linear homogeneous set of differential equations, whose initial state is the only variable affecting its evolution in time, we shall refer to this regulator design problem as the *autonomous regulator problem*. In order to make the comparison between the standard techniques based on transfer functions and the geometric techniques easier, a brief review of the SISO case is first presented.

3.1   The SISO Autonomous Regulator Problem

Let us refer to the closed-loop control system shown in Fig. 1, that includes a plant Σ_p, a regulator Σ_r and two exosystems Σ_{e1} and Σ_{e2}, generating the reference input r, to be tracked by the output y, and the disturbance input d, to be rejected at y, respectively. In this scheme the error e, whose zero setting detects efficiency of tracking and regulation, is the only input of the regulator, while the regulator output coincides with the manipulated input u of the plant. The transfer functions of

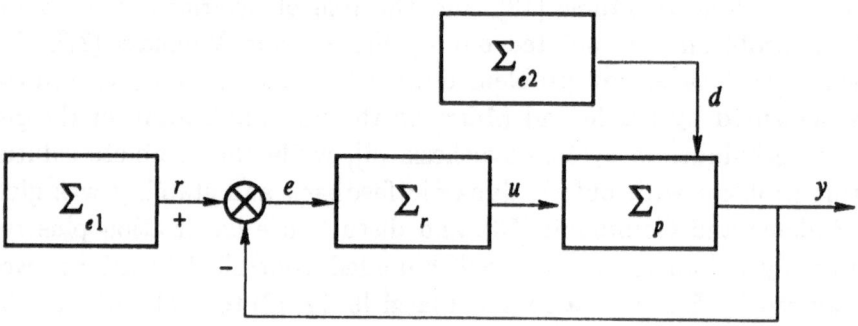

Figure 1: The standard closed-loop control system considered.

the plant (given) and the regulator (to be determined) are

$$G_p(s) = \frac{P_p(s)}{Q_p(s)} \tag{3.1}$$

and

$$G_r(s) = \frac{P_r(s)}{Q_r(s)} \tag{3.2}$$

The polynomials $P_p(s)$, with degree m_p, and $Q_p(s)$, with degree $n_p \geq m_p$, are assumed to be relatively prime. The unknown polynomials $P_r(s)$, with degree m_r, and $Q_r(s)$, with degree n_r, must be such that $n_r \geq m_r$, in order to represent a causal system. The exosystems generate standard test signals, like steps, ramps, parabolas, sinusoids, possibly mixed. Let us suppose that the regulator is disconnected and denote by

$$\frac{P_e(s)}{Q_e(s)} \tag{3.3}$$

a strictly proper transfer function, with $P_e(s)$ having degree n_e, whose impulse response contains all the exosystem modes affecting e, that are assumed to be all not strictly stable. Clearly these modes are represented by the roots of $Q_e(s)$, thus all having non-negative real parts. To achieve asymptotic tracking and disturbance rejection it is necessary that the same modes are generated in the regulator and, possibly, in the plant, thus cancelling those of the exosystem at e. This artifice is called the *internal model principle*. Presence of the internal model and asymptotic stability of the regulation loop (with the exosystem disconnected) are the means to achieve asymptotic regulation and disturbance rejection.

The SISO autonomous regulator problem is: derive, if possible, a causal regulator $G_r(s)$ such that the closed-loop system of Fig. 1 is stable and $\lim_{t \to \infty} e(t) = 0$ for all initial states of the exosystem, i.e., for all $P_e(s)$ in (3.2). In more formal mathematical terms it is stated as follows.

Problem 3.1. (The SISO Autonomous Regulator Problem). Given the transfer function $G_p(s) = P_p(s)/Q_p(s)$, and the polynomial $Q_e'(s)$ such that $Q_p(s) Q_e'(s)$ is divisible by $Q_e(s)$, find, if possible, polynomials $P_r(s)$ and $Q_r(s)$ defining a regulator such that

1. the degree of $P_r(s)$ is not greater than the degree of $Q_r(s)$;

2. $Q_r(s)$ is divisible by $Q_e'(s)$;

3. all the roots of $Q_{cl}(s) = P_r(s)\,P_p(s) + Q_r(s)\,Q_p(s)$ are strictly stable.

Condition 1 expresses causality of the regulator, 2 the internal model principle and 3 strict stability of the control loop. The following theorem holds.

Theorem 3.1. (Solvability of the SISO Autonomous Regulator Problem). Problem 3.1 is solvable with $Q_r(s)$ having degree $n_r \geq n_p + n'_e - 1$ and all the roots of $Q_{cl}(s)$ arbitrary (hence, in particular, strictly stable) if and only if $Q_e(s)$ and $P_p(s)$ are relatively prime, i.e., if and only if all the zeros of the plant are different from the poles of the exosystem.[7]

A very straightforward proof of Theorem 3.1, well known in the literature, is based on some properties of the Diophantine equation

$$A(s)\,X(s) + B(s)\,Y(s) = C(s) \tag{3.4}$$

that is equivalent to a set of linear equations whose unknown quantities are the coefficients of $X(s)$ and $Y(s)$ and known parameters those of $A(s)$, $B(s)$ and $C(s)$. These properties are recalled in the following statement.

Property 3.1. Let us denote by n, m and ℓ the degrees of the polynomials $A(s)$, $B(s)$ and $C(s)$ of equation (3.4). The equation is solvable with respect to $X(s)$ and $Y(s)$ if

- $A(s)$ and $B(s)$ are relatively prime; $\tag{3.5}$
- $\ell \geq m + n$; $\tag{3.6}$

if (3.5) and (3.6) hold, the unique solution of (3.4) $X(s)$, $Y(s)$ satisfies

- the degree of $X(s)$ is $k = \ell - n$; $\tag{3.7}$
- the degree of $Y(s)$ is $h = n - 1$. $\tag{3.8}$

Proof of Theorem 3.1. Let us consider the equation

$$P_r(s)\,P_p(s) + Q'_e(s)\,Q'_r(s)\,Q_p(s) = Q_{cl}(s) \tag{3.9}$$

whose member on the left is the characteristic polynomial of the closed-loop system considered. It is a Diophantine equation of the type (3.4) with $B(s) := P_p(s)$, $A(s) := Q'_e(s)\,Q_p(s)$, $C(s) := Q_{cl}(s)$. The plant being minimal and the assumption on the zeros of the plant ensure condition

---

[7]Augmenting the multiplicity of the internal model poles to compensate for a cancellation with zeros of the plant is not technically acceptable since produces unbounded u.

(3.5). Let $Q_{cl}(s)$ be a polynomial having degree $\ell = 2n_p + n'_e - 1$ with arbitrary roots. In this case the degrees of $Y(s) = P_r(s)$ and $X(s) = Q'_r(s)$ are $m = m_p$, $n = n_p + n'_e$, so that relations (3.7) and (3.8) give $n'_r = n_p - 1$, hence $n_r = n_p + n'_e - 1$, and $m_r = n_r$. A greater value of ℓ simply produces n_r to be increased by the same amount.[8] □

It is worth noting that the internal model produces zeros equal to the exosystem poles in the transfer functions from r to e and from d to e, thus making asymptotic zero setting of e possible. In fact, let us assume $Q_r(s) = Q'_e(s) Q'_r(s)$. This follows from

$$\frac{E(s)}{R(s)} = \frac{1}{1 + G_r(s) G_p(s)} = \frac{Q'_e(s) Q'_r(s) Q_p(s)}{P_r(s) P_p(s) + Q'_e(s) Q'_r(s) Q_p(s)} \quad (3.10)$$

since $Q_e(s)$ is a divisor of $Q'_e(s) Q_p(s)$ by assumption.

Algorithm 3.1. (Solution of the Diophantine Equation). Although the solution of the Diophantine equation is treated in many textbooks, a very concise outline of it, consistent with the previous arguments and symbols, is repeated here for the reader's convenience. Polynomials $A(s)$, $C(s)$ and $X(s)$ are assumed to be monic. Let

$$\begin{aligned}
A(s) &= s^n + a_{n-1} s^{n-1} + \ldots + a_0 & (3.11) \\
B(s) &= b_m s^m + b_{m-1} s^{m-1} + \ldots + b_0 & (3.12) \\
C(s) &= s^\ell + c_{\ell-1} s^{\ell-1} + \ldots + c_0 & (3.13) \\
X(s) &= s^k + x_{k-1} s^{k-1} + \ldots + x_0 & (3.14) \\
Y(s) &= y_h s^h + y_{h-1} s^{h-1} + \ldots + y_0 & (3.15)
\end{aligned}$$

be the given polynomials satisfying the conditions stated in Property 3.1. To maintain notation within acceptably simple terms we refer to the particular case $n=3$, $m-2$, $\ell=8$, hence $h=2$, $k=5$. In this particular

---

[8] A more general result, directly obtainable from the Diophantine equation, is stated as follows: a regulator with n_f fixed poles and m_f fixed zeros while preserving assignability of all the closed-loop poles, has the minimal order $n_r = m_r = n_p + n_f + m_f - 1$, while the number of closed-loop poles is $\ell = 2n_p + n_f + m_f - 1$.

case the set of linear equations equivalent to (3.4) appears in the form

$$
\begin{bmatrix}
1 & 0 & 0 & 0 & 0 & 0 & 0 & 0 \\
a_2 & 1 & 0 & 0 & 0 & 0 & 0 & 0 \\
a_1 & a_2 & 1 & 0 & 0 & 0 & 0 & 0 \\
a_0 & a_1 & a_2 & 1 & 0 & b_2 & 0 & 0 \\
0 & a_0 & a_1 & a_2 & 1 & b_1 & b_2 & 0 \\
0 & 0 & a_0 & a_1 & a_2 & b_0 & b_1 & b_2 \\
0 & 0 & 0 & a_0 & a_1 & 0 & b_0 & b_1 \\
0 & 0 & 0 & 0 & a_0 & 0 & 0 & b_0
\end{bmatrix}
\begin{bmatrix}
x_4 \\ x_3 \\ x_2 \\ x_1 \\ x_0 \\ y_2 \\ y_1 \\ y_0
\end{bmatrix}
=
\begin{bmatrix}
c_7 - a_2 \\ c_6 - a_1 \\ c_5 - a_0 \\ c_4 \\ c_3 \\ c_2 \\ c_1 \\ c_0
\end{bmatrix}
\tag{3.16}
$$

Note that the system matrix is $\ell \times \ell$, with the first k columns reporting the coefficients of $A(s)$ in a band as shown and the last $\ell - k$ columns those of $B(s)$ in a band justified to foot; in the first n elements of the column matrix at the right the coefficients of $A(s)$ from the second up are subtracted from those of $C(s)$ as shown.

3.2 Robustness of regulation in the SISO case

The asymptotic tracking and disturbance rejection property obtained by using a regulator including an accurate and robust replica of the exosystem eigenstructure has the remarkable property of being robust with respect to plant parameter variations so long as closed-loop stability is maintained. The reason for this can be clearly pointed out by analyzing the observability property of a state space realization of the SISO system considered in the previous subsection. Let (A_1, B_1, C_1), (N, M, L, K) be the state space realizations of the plant and the regulator, A_2 be a matrix having the structure of the overall exosystem, i.e., reproducing all the modes of the exosystem Σ_{e1} and Σ_{e2} observable at e with the regulator disconnected. Referring again to Fig. 1, we use for the *regulated system* (the plant plus the exosystems) the following state space representation

$$
\begin{aligned}
\dot{x}(t) &= A\,x(t) + B\,u(t) \\
e(t) &= E\,x(t)
\end{aligned}
\tag{3.17}
$$

with

$$
x := \begin{bmatrix} x_1 \\ x_2 \end{bmatrix}, \quad
A := \begin{bmatrix} A_1 & A_3 \\ O & A_2 \end{bmatrix}, \quad
B := \begin{bmatrix} B_1 \\ O \end{bmatrix}, \quad
E := [E_1 \ \ E_2] \tag{3.18}
$$

Matrices A_1, B_1 and C_1 are assumed to be $n_1 \times n_1$, $n_1 \times 1$ and $1 \times n_1$ respectively, while A_2, that models the exosystem eigenstructure, is assumed to be $n_2 \times n_2$. A_3, $n_1 \times m$, represents the influence of the exosystem Σ_{e2} on the plant, while E_1 and E_2 are set equal to $-C_1$ and C_2 respectively, where C_2, $1 \times n_2$, is the output matrix of A_2 producing the reference r. The regulator equations are

$$
\begin{aligned}
\dot{z}(t) &= N\,z(t) + M\,e(t) \\
u(t) &= L\,z(t) + K\,e(t)
\end{aligned}
\tag{3.19}
$$

where N, M, L are $m \times m$, $m \times 1$ and $1 \times m$ respectively, while K is a scalar. Consistency with the previous transfer function description implies $n = n_p$, $m = n_r$ and $n_2 = n_e$.

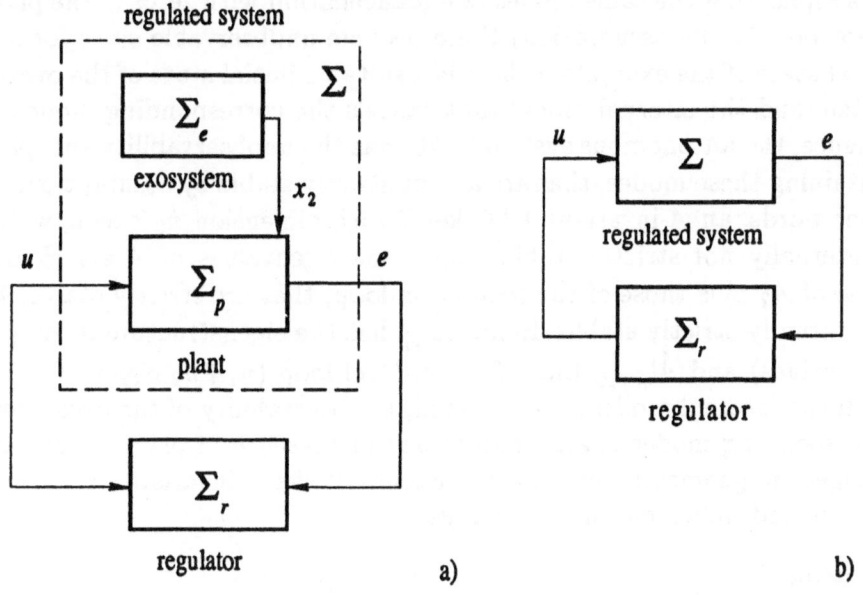

Figure 2: Block diagrams referring to the state-space representation of a MIMO system.

Fig. 2,a shows the standard connection of the exosystem, plant and regulator, while Fig. 2,b points out the essence of the autonomous regulator problem: the regulated system Σ (given) with its closed-loop connection with the regulator Σ_r (to be determined). The overall system is described as the autonomous *extended system*

$$
\begin{aligned}
\dot{\hat{x}}(t) &= \hat{A}\,\hat{x}(t) \\
e(t) &= \hat{E}\,\hat{x}(t)
\end{aligned}
\tag{3.20}
$$

with

$$\hat{A} := \begin{bmatrix} A_1 + B_1KE_1 & A_3 + B_1KE_2 & B_1L \\ O & A_2 & O \\ ME_1 & ME_2 & N \end{bmatrix} , \quad \hat{x} := \begin{bmatrix} x_1 \\ x_2 \\ z \end{bmatrix} ,$$

$$\hat{E} := [E_1 \quad E_2 \quad O] \tag{3.21}$$

Referring to (3.20), the asymptotic tracking and regulation property is explained in geometric terms as follows: assume that the pair (E, A) is observable; this implies, in particular, that the exosystem modes may all affect e. If a regulator with the internal model is connected as shown, thus replicating the same modes at e (cancellation with zeros of the plant is not possible by assumption) these become unobservable since for any initial state of the exosystem there is a suitable initial state of the overall system and the internal model that cancels the corresponding modes at e; hence, the autonomous system (3.20) has the unobservability subspace containing these modes, that are all not strictly stable by assumption. In other words, an $\hat{A}$-invariant $\hat{\mathcal{W}} \subseteq \ker \hat{E}$ with dimension n_2 exists, which is internally not strictly stable. Since the eigenvalues of $\hat{A}$ are clearly those of A_2 plus those of the regulation loop, that are strictly stable, $\hat{\mathcal{W}}$ is externally strictly stable. Hence $\hat{A}|_{\hat{\mathcal{W}}}$ has the eigenstructure of A_2 (n_2 eigenvalues) and $\hat{A}|_{\hat{\mathcal{X}}/\hat{\mathcal{W}}}$ that of control loop ($n_1 + n_2$ eigenvalues).

If the internal model is maintained, unobservability of the exosystem corresponding modes is also maintained in presence of plant parameter changes. In geometric terms, the existence of the $\hat{A}$-invariant $\hat{\mathcal{W}} \subseteq \ker \hat{E}$ is preserved under parameter changes.

Remarks:

1. Robustness of the internal model is easily achievable when its eigenvalues are all zero, i.e., when the signals to be asymptotically tracked or rejected are linear combinations of steps, ramps, parabolas, etc.

2. Since the closed-loop poles are all assignable, the modes of the error variable are arbitrary.

3. The invariant zeros of (A, B, E) are those of the plant (A_1, B_1, C_1) plus the eigenvalues of A_2.

4. It is possible that the poles of the overall exosystem are a subset of those of the internal model (for instance, a double pole at the

origin in the internal model produces steady-state regulation also with an exosystem having a single pole at the origin).

3.3 The MIMO Autonomous Regulator Problem

Let us consider again equations (3.17–3.19) and refer them to the MIMO case, by assumimg that matrices B_1 and C_1 are now $n_1 \times p$ and $q \times n_1$, while M, L, K and C_2 are $m \times q$, $p \times m$, $p \times q$ and $q \times m$ respectively. We also assume that (A_1, B_1) is reachable and (E, A) observable.[9]

In the corresponding state space $\mathcal{X}$ with dimension $n := n_1 + n_2$ we define the subspace *plant* $\mathcal{P}$ through

$$\mathcal{P} := \{\, x \,:\, x_2 = 0 \,\} = \text{im} \begin{bmatrix} I_{n_1} \\ O \end{bmatrix} \tag{3.22}$$

Similarly, in the extended state space $\hat{\mathcal{X}}$ with dimension $\hat{n} := n_1 + n_2 + m$ the *extended plant* $\hat{\mathcal{P}}$ is defined as

$$\hat{\mathcal{P}} := \{\, \hat{x} \,:\, x_2 = 0 \,\} = \text{im} \begin{bmatrix} I_{n_1} & O \\ O & O \\ O & I_m \end{bmatrix} \tag{3.23}$$

Clearly $\mathcal{P}$ and $\hat{\mathcal{P}}$ are an A-invariant and an $\hat{A}$-invariant respectively.

The MIMO autonomous regulator problem is: derive, if possible, a regulator (N, M, L, K) such that the closed-loop system with the exosystem disconnected is stable and $\lim_{t \to \infty} e(t) = 0$ for all initial states of the autonomous extended system. In geometric terms it is stated as follows.

Problem 3.2. (The MIMO Autonomous Regulator Problem). Let us refer to the extended system (3.20) and assume $\hat{\mathcal{E}} := \ker \hat{E}$. Given the mathematical model (3.17, 3.18) of the plant and the exosystem determine, if possible, a regulator (N, M, L, K) such that an $\hat{A}$-invariant $\hat{\mathcal{L}}$ exists satisfying

1. $\hat{\mathcal{L}} \subseteq \hat{\mathcal{E}}$;

2. $\sigma(\hat{A}|_{\hat{\mathcal{X}}/\hat{\mathcal{L}}}) \subseteq \mathbf{C}_-$.

---

[9]Relaxing these assumptions to (A_1, B_1) being stabilizable and (E, A) detectable does not significantly affect the solution: the only consequence is that the error dynamics is no longer arbitrary.

Lemma 3.1. Problem 3.2 has a solution if and only if an $\hat{A}$-invariant $\hat{\mathcal{W}}$ exists such that

$$\hat{\mathcal{W}} \subseteq \hat{\mathcal{E}} \tag{3.24}$$

$$\hat{\mathcal{W}} \oplus \hat{\mathcal{P}} = \hat{\mathcal{X}} \tag{3.25}$$

$$\sigma(\hat{A}|_{\hat{\mathcal{X}}/\hat{\mathcal{W}}}) \subseteq \mathbf{C}_- \tag{3.26}$$

Proof. Only if. Suppose that Problem 3.2 has a solution and let $\hat{\mathcal{W}}^+$ be the subspace of the non-strictly stable modes of $\hat{A}$, that is an $\hat{A}$-invariant. Since the only non-strictly stable eigenvalues are those of A_2, while the other eigenvalues, strictly stable, are those of the regulation loop, i.e., those of $\hat{A}|_{\hat{\mathcal{X}}/\hat{\mathcal{P}}}$, the dimension of $\hat{\mathcal{W}}^+$ is n_2. Hence, by the requirement 2 of Problem 3.2, $\hat{\mathcal{W}}^+ \subseteq \hat{\mathcal{L}}$. It follows that $\hat{\mathcal{W}}^+$ is the minimum $\hat{A}$-invariant meeting the geometric conditions of Problem 3.2.

If. Assume $\hat{\mathcal{L}} := \hat{\mathcal{W}}^+$ in the statement of Problem 3.2.   □

Owing to (3.25) $\hat{\mathcal{W}}$ can be expressed as the image of a basis matrix with a particular structure, i.e.,

$$\hat{\mathcal{W}} = \mathrm{im}\hat{W} \quad \text{with} \quad \hat{W} = \begin{bmatrix} X_1 \\ I_{n_2} \\ Z \end{bmatrix} \tag{3.27}$$

The main Theorem on asymptotic regulation in the MIMO case is stated in strictly geometric terms as follows.

Theorem 3.2. (Solvability of the MIMO Autonomous Regulator Problem: non-constructive necessary and sufficient conditions). Let us assume $\mathcal{E} := \ker E$. Problem 3.2 admits a solution if and only if an (A, B)-controlled invariant $\mathcal{V}$ exists such that

$$\mathcal{V} \subseteq \mathcal{E} \tag{3.28}$$

$$\mathcal{V} \oplus \mathcal{P} = \mathcal{X} \tag{3.29}$$

Proof. Only if. Let us refer to (3.27), with $\hat{\mathcal{W}}$ satisfying the properties stated in Lemma 3.1 and assume

$$\mathcal{V} = \mathrm{im}V \quad \text{with} \quad V = \begin{bmatrix} X_1 \\ I_{n_2} \end{bmatrix} \tag{3.30}$$

$\mathcal{V}$ is an $(A, \mathrm{im}B)$-controlled invariant. In fact, owing to (2.4) a matrix S exists such that $\hat{A}\hat{W} = \hat{A}S$, or

$$\begin{bmatrix} (A_1 + B_1 K E_1) X_1 + A_3 + B_1 K E_2 + B_1 L Z \\ A_2 \\ M E_1 X_1 + M E_2 + N Z \end{bmatrix} = \begin{bmatrix} X_1 \\ I_{n_2} \\ Z \end{bmatrix} S \tag{3.31}$$

By suitably collecting elements we obtain

$$\begin{bmatrix} A_1 X_1 + A_3 \\ A_2 \end{bmatrix} = \begin{bmatrix} X_1 S - B_1 (K E_1 X_1 + K E_2 + L Z) \\ S \end{bmatrix} \quad (3.32)$$

that can be expressed as $AV = VS + BU$, so that V is an (A, B)-controlled invariant by virtue of (2.13). Due to the particular structure of $\hat{E}$ shown in (3.21), from $\hat{E}\hat{W} = O$ it follows that $EV = O$, hence (3.28) holds. Finally, (3.29) immediately follows from the structure of the basis matrix V.

If. Let V be an (A, B)-controlled invariant satisfying (3.29), hence having a basis matrix V with the structure shown in (3.30): a state-feedback matrix $F = [F_1 \ F_2]$ exists, partitioned according to (3.18), such that $(A + BF)V \subseteq V$ and the eigenvalues of $A_1 + B_1 F_1$ are arbitrarily assigned. In fact, consider the similarity transformation defined as $T = [T_1 \ T_2]$ with $T_1 = [I_{n_1} \ O]^T$, $T_2 = V$. The transformed matrices $A' := T^{-1}AT$, $B' := T^{-1}B$, and $E' := ET$ have the structures

$$A' = \begin{bmatrix} A_1 & A_3' \\ O & A_2 \end{bmatrix}, \quad B' = \begin{bmatrix} B_1 \\ O \end{bmatrix}, \quad E' = [E_1 \ O] \quad (3.33)$$

Since (A_1, B_1) is reachable, a matrix F_1 exists such that $A_1 + B_1 F_1$ has any given set of n_1 eigenvalues; since V is a controlled invariant, a matrix F_2' exists such that $A_3' + B_1 F_2' = O$. F referred to the original basis is computed as $F = [F_1 \ F_2'] T^{-1}$. Furthermore, let G be such that $A + GE$ has any given set of n eigenvalues, which is possible since (E, A) is observable. We claim that the regulator defined by

$$N := A + BF + GE, \quad M := -G, \quad L := F, \quad K := O \quad (3.34)$$

solves Problem 3.2. In fact Lemma 3.1 is satisfied with

$$\hat{A} = \begin{bmatrix} A & BF \\ -GE & A + BF + GE \end{bmatrix}, \quad \hat{E} = [E \ O], \quad \hat{W} = \begin{bmatrix} V \\ V \end{bmatrix} \quad (3.35)$$

as is shown with the similarity transformation defined by

$$T = T^{-1} = \begin{bmatrix} I_n & O \\ I_n & -I_n \end{bmatrix} \quad (3.36)$$

The corresponding matrices $\hat{A}' := T^{-1}\hat{A}T$, $\hat{E}' := \hat{E}T$ and $\hat{W}' := T^{-1}\hat{W}$ appear in the forms

$$\hat{A}' = \begin{bmatrix} A + BF & -BF \\ O & A + GE \end{bmatrix}, \quad \hat{E}' = [E \ O], \quad \hat{W}' = \begin{bmatrix} V \\ O \end{bmatrix} \quad (3.37)$$

that allow easy checking of conditions stated in Lemma 3.1, by also using the previously defined partitioning of the involved submatrices.   □

Conditions stated in Theorem 3.2, although necessary and sufficient, are non-constructive, since the existence of an controlled invariant satisfying (3.28, 3.29), that is here called the *resolvent*, is not easy to check. However, the "if" part of the proof provides an algorithm to derive a controller with the internal model of the exosystem when the resolvent is given. The order of the controller is $n = n_1 + n_2$ (that of the plant plus that of the exosystem) like in the SISO case. The following theorem gives conditions similar to those of Theorem 3.1 for the SISO case that are also only sufficient, but very important in practice, since avoid unboundedness of the manipulated input u due to pole-zero cancellation between the plant and the regulator.

Theorem 3.3. (Solvability of the MIMO Autonomous Regulator Problem: constructive sufficient and almost necessary[10] conditions). Let us assume $V^* := \max V(A, B, \mathcal{E})$. Problem 3.2 admits a solution if

$$V^* + \mathcal{P} = \mathcal{X} \tag{3.38}$$

$$\mathcal{Z}(A_1, B_1, E_1) \cap \sigma(A_2) = \emptyset \tag{3.39}$$

Proof. We show that the stated conditions allow computation of a resolvent. Let F be a matrix such that $(A+BF)V^* \subseteq V^*$ and introduce the similarity transformation $T := [T_1 \ T_2 \ T_3]$, with $\text{im} T_1 = V^* \cap \mathcal{P}$, $\text{im} T_2 = V^*$ and T_3 such that T is nonsingular. Recall that $\mathcal{P}$ is an A-invariant and note that, owing to the particular structure of B, it also is an $(A+BF)$-invariant for any F. In the new basis the linear transformation $A+BF$ has the structure

$$A' = T^{-1}(A+BF)\, T = \begin{bmatrix} A'_{11} & A'_{12} & A'_{13} \\ O & A'_{22} & O \\ O & O & A'_{33} \end{bmatrix} \tag{3.40}$$

By a dimensionality argument the eigenvalues of the exosystem are those of A'_{22}, while the invariant zeros of (A_1, B_1, E_1) are a subset of $\sigma(A'_{11})$ since $\mathcal{R}_{V^*}$ is contained in $V^* \cap \mathcal{P}$. All the other elements of $\sigma(A'_{11})$ are arbitrarily assignable with F. Hence, owing to (39), the Sylvester equation

$$A'_{11} X - X A'_{22} = -A'_{12} \tag{3.41}$$

---

[10]The conditions are necessary if the control variable u must remain bounded: like in the SISO case, this is possible also when the output y is unbounded if a part of the internal model is contained in the plant.

admits a unique solution. The matrix $V := T_1 X + T_2$ is a basis matrix of an (A, B)-controlled invariant V satisfying (3.28), (3.29). □

If $V^* \cap B = \{0\}$, matrix A'_{11} is independent of F. Thus the following corollary is also proved.

Corollary 3.1. (Uniqueness of the resolvent). If the plant is invertible and (3.38, 3.39) are satisfied, a unique (A, B)-controlled invariant V satisfying conditions (3.28, 3.29) exists.

Remarks:

1. The proof of Theorem 3.3 provides the computational framework to derive a resolvent when the sufficient conditions stated (that are also necessary if the input of the plant must remain bounded) are satisfied.

2. Relations (3.38) and (3.39) are respectively a *structural condition* and a *spectral condition*; they are easily checkable by means of Algorithms 2.6–2.8 described in the previous section.

3. When a resolvent has been determined by means of the computational procedure described in the proof of Theorem 3.3, it can be used to derive a regulator with the procedure outlined in the "if" part of the proof of Theorem 3.2.

4. The order of the obtained regulator is n (that of the plant plus that of the exosystem) with the corresponding $2n_1 + n_2$ closed-loop eigenvalues completely assignable under the assumption that (A_1, B_1) is reachable and (E, A) observable.

5. The internal model principle is satisfied since the eigenstructure of the regulator system matrix $(A + BF + GE)$ contains that of A_2: in fact, the subspace V, that is an $(A + BF)$-invariant with the internal eigenstructure equal to that of A_2, is also an $(A + BF + GE)$-invariant with the same internal eigenstructure because it is contained in $\ker E$. Thus, if we consider the connection from the exosystem to the overall system as the input of the feedback loop and the error variable e as the output, the eigenstructure of A_2 is a part of the invariant zero structure.

3.4 Robustness of regulation in the MIMO case

Unfortunately, having the eigenstructure of the exosystem as a part of the invariant zero structure of the feedback loop is not sufficient to obtain

robustness of regulation in presence of small parameter changes also in
the MIMO case. The reason for this can be explained in simple terms
as follows: an invariant zero of a generic triple (A, B, C) is, according
to Definition 2.9, an internal unassignable eigenvalue of the maximal
(A, B) controlled invariant contained in C and if the corresponding mode
is *suitably* injected at the input, an initial state nulling the output exists.
In the multivariable regulation loop derived in the proof of Theorem 3.2
the internal model actually nulls the modes of the exosystem appearing
at e, since it produces corresponding zeros in the closed-loop system, but
only if they are injected in the loop in the precise way represented by the
mathematical model (A, B, E) assumed: if this is subject to parameter
variations, since the influence of each mode on each output is subject to
change unpredictibly and independently of those of the corresponding
mode generated in the regulator, asymptotic perfect zero setting of the
error is no longer possible.

A possible remedy for this is to repeat the exosystem internal model
in the regulator once for each controlled output, thus making automatic
compensation of any parameter drift possible, if it is small enough to
preserve the strict stability of the regulation loop. We shall show with
a constructive argument that such an asymptotically robust regulator
exists if the conditions stated in Theorem 3.2 are satified, without any
additional requirement.

Algorithm 3.2. (The Francis Multivariable Asymptotically Robust
Regulator). Assume as the matrix of the internal model

$$A_{2e} := \begin{bmatrix} A_2 & O & \dots & O \\ O & A_2 & \dots & O \\ \vdots & \vdots & \ddots & \vdots \\ O & O & \dots & A_2 \end{bmatrix} \tag{3.42}$$

where A_2 is replicated as many times as there are regulated output
components. Let n_{2e} be the dimension of A_{2e}. Then define the extension
of the controlled system

$$A_e := \begin{bmatrix} A_1 & A_{3e} \\ O & A_{2e} \end{bmatrix}, \quad B_e := \begin{bmatrix} B_1 \\ O \end{bmatrix},$$

$$E_e := [E_1 \quad O], \quad P_e := \begin{bmatrix} I_{n_1} \\ O \end{bmatrix} \tag{3.43}$$

where A_{3e} has been determined under the only requirement that (E_e, A_e)
is observable. The existence of any (A, B)-controlled invariant $\mathcal{V} \subseteq \mathcal{E}$ satisfying $\mathcal{V} + \mathcal{P} = \mathcal{X}$ implies that B_1 is such that every mode of the exosystem

appearing at e (from which the exosystem is completely observable) can be cancelled by a suitable feedback from the state of the exosystem itself to u. Since this is an intrinsic property of (A_1, B_1, E_1), it is also valid for the new system (3.43) i.e., $\mathcal{V}^* + \mathcal{P} = \mathcal{X}$ implies $\mathcal{V}_e^* + \mathcal{P}_e = \mathcal{X}_e$ with $\mathcal{V}_e^* := \max \mathcal{V}(A_e, B_e, \mathcal{E}_e)$ $(\mathcal{E}_e := \ker E_e)$. Since the conditions of Theorem 3.3 are also valid for $\mathcal{V}_e^*$, an (A_e, B_e)-controlled invariant $\mathcal{V}_e$ exists such that $\mathcal{V}_e \oplus \mathcal{P}_e = \mathcal{X}_e$. Determine $F_e = [F_1 \ F_{2e}]$ such that $(A_e + B_e F_e) \mathcal{V}_e \subseteq \mathcal{V}_e$ and $A_1 + B_1 F_1$ is stable, G_e such that $A_e + G_e E_e$ is stable, and assume

$$N := A_e + B_e F_e + G_e E_e, \quad M := -G_e, \quad L := F_e, \quad K := O \quad (3.44)$$

With this assumption the conditions stated in Lemma 3.1 hold. In fact, from

$$\hat{A} = \begin{bmatrix} A_1 & A_3 & B_1 F_e \\ O & A_2 & O \\ -G_e E_1 & -G_e E_2 & A_e + B_e F_e + G_e E_e \end{bmatrix}, \quad (3.45)$$

$$\hat{E} = [E_1 \ E_2 \ O]$$

by means of the similarity transformation

$$T = T^{-1} := \begin{bmatrix} I_{n_1} & O & O \\ O & I_{n_2} & O \\ R & O & -I_{n_e} \end{bmatrix}, \quad \text{with } R := \begin{bmatrix} I_{n_1} \\ O \end{bmatrix} \quad (3.46)$$

we derive as $\hat{A}' := T^{-1} \hat{A} T$ and $\hat{E}' := \hat{E} T$ the matrices

$$\hat{A}' = \begin{bmatrix} A_1 + B_1 F_1 & A_3 & -B_1 F_e \\ O & A_2 & O \\ O & S & A + G_e E \end{bmatrix}, \quad \hat{E}' = [E_1 \ E_2 \ O] \quad (3.47)$$

In (3.46) $n_e := n_1 + n_{2e}$ and R is assumed to be $n_e \times n_1$ while in (3.47) S denotes a nonzero matrix. Since the closed-loop system is obtained from (3.47) by deleting the second row and column, we conclude that the eigenvalues of the regulation loop can be arbitrarily assigned. Owing to the presence of a "robust" internal model, that sets asymptotically to zero all the exosystem modes at every component of e, the pair $(\hat{E}, \hat{A})$ has a "robust" unobservability subspace $\hat{\mathcal{W}}$ with dimension n_2 that can be computed both as $\max \mathcal{L}(\hat{A}, \hat{E})$ or as the unique complement of the $\hat{A}$-invariant $\hat{\mathcal{P}}$.

Remarks:

1. Algorithm 3.2 allows achievement of asymptotic robustness with a regulator of order $n_1 + q \times n_{2e}$, q being the number of the regulated

output components (the components of e, assumed to be as many as the components of the output y of the plant).

2. The order of the regulator can be reduced if for each regulated output e_i an internal model is used, reproducing only the maximal eigenstructure of the exosystem observable at e_i when parameters change. This can be obtained by choosing suitable submatrices in A_{2e} and A_{3e} that, for every eigenvalue of the exosystem, make observable at e_i an eigenstructure belonging to the internal model and the plant equal to the eigenstructure of the exosystem that is observable at e_i.

3. The plant can complete the action of a particular internal model, thus making reduction of its order possible, if some of its eigenvalues are equal to those of the exosystem observable at the corresponding output; however, these eigenvalues of the plant must be suitably shared between outputs, avoiding repetitions that may cause lack of robustness of regulation.

Notes and References. In many textbooks on automatic control systems, pole assignment is solved, even in the SISO case, by converting the system transfer function into a quadruple (A, B, C, D) and using state feedback through a full-order or a reduced-order observer. The reason for this is probably only a historical habit, since resort to the Diophantine equation appears to be more convenient, at least in the SISO case, since it gives the result directly in polynomial form. It also makes it possible to introduce any given transfer function as a part of the regulator. Some well known books that use the Diophantine equation are those by Kučera [34], Åström and Wittenmark [1, 2], Middleton and Goodwin [41] and Chen [15] (this also in the MIMO case). The properties of the exosystem's internal model in the MIMO case were studied by Francis, Sebakhy and Wonham [26, 28], while the most complete solution to the multivariable regulation problem with multiple internal model and complete eigenvalue assignment is due to Francis [24, 25]: this solution has been simply translated into geometric terms in the previously reported algorithm. Of course, eigenvalue assignability with the internal model solves the regulator problem from a strictly mathematical viewpoint, but leaves an important practical problem unsolved: robust stability. A significant recent contribution to the regulator problem with robust stability, stated in the same geometric terms that are used herein, has been made by Cevik and Schumacher [14].

4 Perfect Tracking with Feedforward, Preview and Preaction

This section completes the geometric approach to regulation theory by adding the second degree of freedom: feedforward. The object is to complete the good asymptotic performance of the feedback loop with a satisfactory transient behavior. In the standard approach this is obtained through a suitable choice of a certain number of zeros of the overall transfer function from reference to output, that are freely assignable without any influence on the closed-loop poles. When the plant has some finite delay or is nonminimum-phase the standard causal approach is no longer satisfactory, and resort to special controllers based on preview of the signals to be tracked, is necessary.

4.1 Model-Reference SISO Continuous-Time Systems

Let us consider the regulation scheme shown in Fig. 3,a. Note that there is a substantial difference between the regulator of Fig. 1, that implements only the closed-loop stability and asymptotic tracking features, and that of Fig. 3,a, that must also give a good response to the input r, or, possibly, reproduce it with negligible error: the new regulator has two inputs with independent sets of zeros.

To preserve the internal model, the regulator is realized as shown in Fig. 3,b, i.e., by interconnection of an *asymptotic tracking unit* Σ_{ra}, reproducing only the internal model, and a *stabilizing unit* Σ_{rs}, that assign all the closed-loop poles. Let

$$\frac{P_{ra}(s)}{Q_{ra}(s)}, \quad \frac{P_{rsv}(s)}{Q_{rs}(s)}, \quad \frac{P_{rsy}(s)}{Q_{rs}(s)}, \quad \frac{P_p(s)}{Q_p(s)} \tag{4.1}$$

be the transfer functions of the asymptotic tracking unit, the stabilizing unit with respect to input v, the stabilizing unit with respect to input y and the regulated plant respectively. Note that the stabilizing unit has a unique dynamics (hence a unique state-space realization, with two different input distribution matrices).

$Q_{ra}(s)$ depends on the exosystem to track, hence is assumed to be a-priori known, with degree n'_e.[11] Let m_p and n_p be the degrees of $P_p(s)$

---

[11] We denote, like in the previous section, by n_e the degree of the exosystem and by $n'_e \leq n_e$ that of the internal model, that is less in the case where the plant has some poles equal to those of the exosystem.

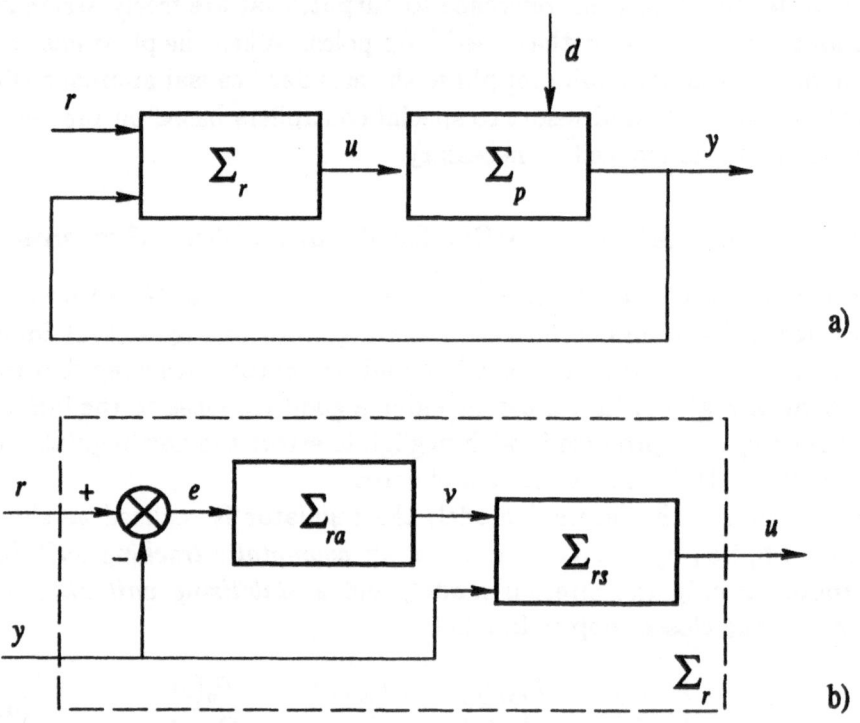

Figure 3: The reference scheme for the two–degrees–of–freedom regulation of a continuous–time systems.

and $Q_p(s)$ respectively; the degree of $Q_{rs}(s)$ is n_p, if the stabilizing unit is realized through a full-order observer, or $n_p - 1$, if a reduced-order observer is used. Polynomials $Q_{rs}(s)$ and $P_{rsy}(s)$ are determined by assigning the $2n_p$ or $2n_p - 1$ poles of the inner loop, while $P_{ra}(s), P_{rsv}(s)$ are free, with degrees ranging from 0 to n'_e and from 0 to n_p or $n_p - 1$ respectively. They can be used to implement a *model-reference* design, where the transfer function $G_0(s) = P_0(s)/Q_0(s)$ from r to y is arbitrarily assigned (with only some generic and easy-to-handle constraints) if the plant is minimum-phase, i.e., with all its zeros stable. This is set in more precise terms in the following Theorem.

Theorem 4.1. (Constraints on the Model-Reference Transfer Function). Let us assume that the plant is minimum-phase. The regulator shown in Fig. 3,b allows reproduction of a model-reference transfer function $P_0(s)/Q_0(s)$ with $P_0(s)$ having degree m_0 and $Q_0(s)$ degree n_0 under the constraints:

- $Q_{ra}(s)$ is a divisor of $Q_0(s) - P_0(s)$; $\qquad$ (4.2)
- $m_0 \geq n'_e - 1$; $\qquad$ (4.3)
- $n_0 \geq m_0 + n_p - m_p$. $\qquad$ (4.4)

Proof. Let

$$H(s) := \frac{P_{rsv}(s) \, P_p(s)}{Q_{cl}(s)} \quad \text{and} \quad H_1 := \frac{P_{ra}(s)}{Q_{ra}(s)} \, H(s) \qquad (4.5)$$

be the transfer functions of the stabilized plant and the outer loop respectively. Condition (4.2) comes from

$$\frac{P_0(s)}{Q_0(s)} = \frac{H_1(s)}{1 + H_1(s)} \quad \text{hence} \quad H_1(s) = \frac{P_0(s)}{Q_0(s) - P_0(s)} \qquad (4.6)$$

By factorization of $P_0(s)$ in (4.2) it follows that

$$Q_0(s) = Q_{ra}(s) \, Q'_0(s) + P'_0(s) \, P''_0(s) \qquad (4.7)$$

that can be interpreted as a Diophantine equation with $C(s) := Q_0(s)$, $A(s) := Q_{ra}(s)$ and $B(s) := P''_0(s)$ (given) and $X(s) := Q'_0(s)$, $Y(s) := P'_0(s)$ (to be determined). Denote by ℓ the degree of $Q_0(s)$ and ℓ_1 that of $P''_0(s)$, that is subject to the constraint $\ell_1 \leq \ell - n'_e$: it follows that the degree of $Q'_0(s)$ is $\ell - n'_e$ and the degree of $P'_0(s)$ is $n'_e - 1$. If we assume $Q''_0(s)$ with minimum degree, i.e., equal to a real constant, (4.3) immediately follows. Relations on the right of (4.5) and (4.6) imply

$$\frac{P_{ra}(s)}{Q_{ra}(s)} = \frac{P_0(s)}{Q_0(s) - P_0(s)} \frac{Q_{cl}(s)}{P_{rsv}(s) P_p(s)} \tag{4.8}$$

from which it is easily seen that, if polynomials $P_{rsv}(s)$ and $Q_{cl}(s)$ are assumed to perform the maximum number of pole-zero cancellations, (4.4) holds with the equality sign.   □

If the plant is nonminumum-phase its unstable zeros cannot be cancelled with corresponding poles in the regulator, so they must be repeated in the model as roots of $P_0(s)$, thus seriously conditioning the overall system behavior and making the model reference design questionable. In this case the conditions stated in Theorem 4.1 still hold, provided, of course, that in the member on the right of (4.3) the number of the unstable zeros of the plant is also considered.

It is worth noting that the stabilizing unit Σ_s may receive more than one feedback signal from the plant to improve robustness of the stabilizing inner loop; in this case the unit uses a reduced-order observer: Property 4.1 still holds, since the numbers of the assignable zeros and the assignable poles of $H(s)$ are reduced by the same amount.

4.2 Preview and Preaction in SISO Discrete-Time Systems

In the discrete-time case things go significantly better. The reason for this is not inherent in the type of system, but in the mind of control engineers and scientists, who usually accept delays in tracking in the discrete-time case and do not in the continuous-time case. This is a very crucial point because delayed tracking implies preview, hence renouncing causality. Knowing the signal to be tracked a certain time in advance is called *preview* and acting on the manipulated input of the plant to prepare tracking is called *preaction*. Preaction time is not necessarily equal to preview time: in fact, in some cases tracking with zero error is possible without any preaction. Of course, preaction is not possible without preview and for any input segment to be tracked preaction time must be less than or at most equal to preview time.

It has been shown in the literature that preview and consequent preaction provide substantial benefits in systems with delays (this is obvious) and in nonminimum-phase systems (this may be less apparent). As far as the scenario where preview and preaction give benefits in tracking problems is concerned, it is worth distinguishing the following three cases:

1. infinite preview time and arbitrarily large preaction time (tracking a profile with a machine-tool);

2. preview and preaction times variable during operation (tracking a route with an aircraft);

3. preview and preaction times fixed during operation (like in receding-horizon model-predictive control).

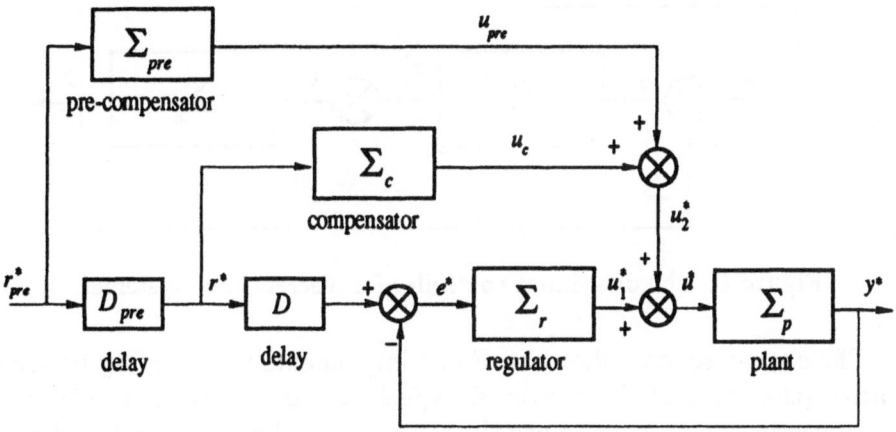

Figure 4: The reference scheme for the two-degrees-of-freedom regulation of discrete-time systems.

In the discrete-time case the objective of control is usually more ambitious: model-reference with $G_0(z) = 1$, i.e., perfect tracking. The control system in the more general case has the structure shown in Fig. 4: we accept a delay D in tracking, corresponding to the relative degree of the plant, including a possible finite delay, i.e., a multiple pole at $z = 0$. Hence, even if the blocks D_{pre} and Σ_{pre} are absent, there is a preaction through the feedforward compensator Σ_c. This compensator is sufficient to ensure perfect tracking when the plant is minimum-phase. If the plant is nonminimum-phase however a greater preaction time may be required, since in this case the preaction time does not only depend on the plant relative degree, but also on the absolute value of the unstable zeros: in this case a further delay D_{pre} and a special pre-compensator Σ_{pre} are necessary to obtain almost perfect tracking. Perfect tracking, in the strict mathematical sense, is obtained when delay D_{pre} is infinity:

in practice, almost perfect tracking is achieved if the total preaction time corresponding to D_{pre} and D is such that the mode corresponding to the unstable zero with absolute value closest to one, considered in the reverse time direction, becomes negligible with respect to its initial value.

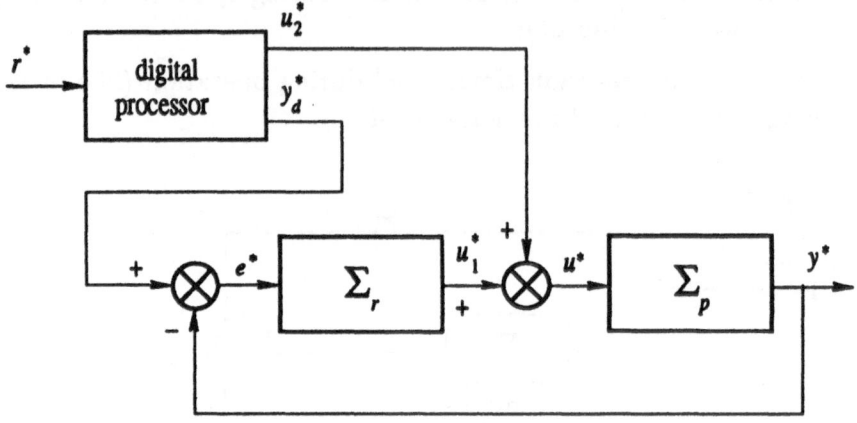

Figure 5: The dual-input controller for discrete-time systems.

The control scheme shown in Fig. 4 implements the receding-horizon control (the third of the previously specified cases). To fit in also the other cases it is necessary to use the more general implementation shown in Fig. 5, where a digital processor, performing simply the convolution of the previewed part of the reference signal to be tracked with a precomputed sequence depending on the plant transfer function, directly generates the feedforward signal.

4.3   Perfect Tracking in SISO Continuous-Time Systems

We now propose a special control that extends to the continuous-time case all the characterizing features of the dual-input controller shown in Fig. 5. Consider the control system structured as in Fig. 6 consisting of a *plant* Σ_p, a *regulator* Σ_r, a *compensator* Σ_c, a *signal generator* (or exosystem) Σ_e and (possibly) a *filter* Σ_f. The control loop is standard, designed to achieve asymptotic robust tracking of the exosystem signals and disturbance rejection. The *supervising unit* shown in the figure receives the reference input r and reproduces it, possibly with some delay, by changing the exosystem state at suitable instants of time; it also can change the states of Σ_r and Σ_c at any time. The filter Σ_f is included in the forward path to obtain a reference signal $y_d(t)$ smooth enough to be tracked by the plant with piecewise-continuous control input $u(t)$.

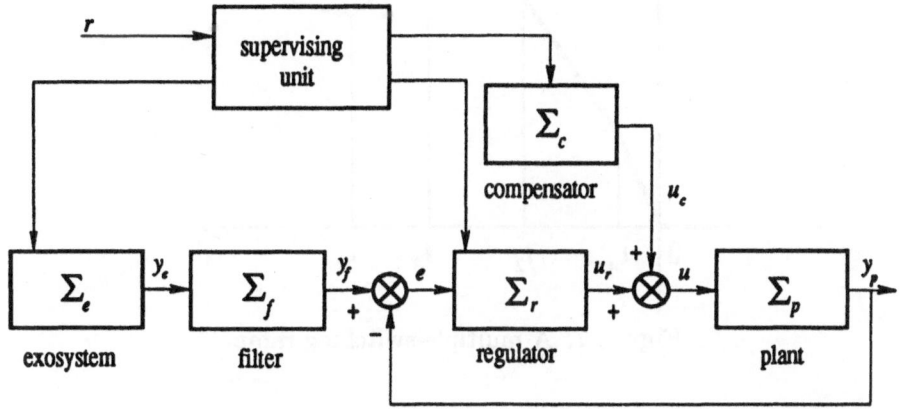

Figure 6: The SISO perfect or almost-perfect tracking implementation.

We assume that the plant, possibly nonminimum-phase, has no zeros on the imaginary axis, and that an internal model of the exosystem is included in the feedforward path of the regulation loop. Hence the loop is of the standard type discussed in the previous section. The transfer functions of the regulator and the plant are denoted by

$$G_r(s) = \frac{P_r(s)}{Q_r(s)}, \quad G_p(s) = \frac{P_p(s)}{Q_p(s)} \tag{4.9}$$

The signal generated by Σ_e is described in terms of its $\mathcal{L}$-transform by

$$Y_e(s) = \frac{\alpha}{s^\ell}. \tag{4.10}$$

A more realistic signal to be tracked is represented by a time-shifted linear combination of the above signal (see Fig. 7), i.e.,

$$Y_e(s) = \frac{\alpha_0}{s^\ell} + \frac{\alpha_1}{s^\ell} e^{-t_1 s} + \ldots + \frac{\alpha_k}{s^\ell} e^{-t_k s}, \tag{4.11}$$

where $0 < t_1 < t_2 < \ldots t_k$, for some integer k. Due to linearity and time-invariance of our system the solution this case is obtained as a time-shifted linear combination of the solution for (4.10). Therefore, without loss of generality we will assume zero as the unique switching time.

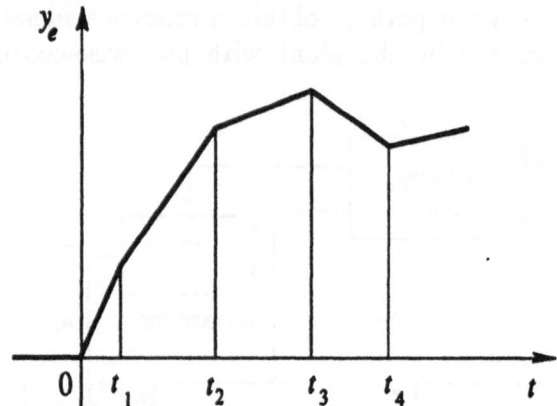

Figure 7: A multiple-switching ramp.

The filter has no zeros and is described by the transfer function

$$G_f(s) = \frac{k_f}{Q_f(s)} \qquad (4.12)$$

where k_f is a constant and $Q_f(s)$, having degree n_f, is assumed to be strictly stable.

Let $Q_e(s) := s^{\ell}$: the presence of the internal model implies $Q_e(s) = Q'_e(s)\,Q''_e(s)$, with $Q'_e(s)$ being a divisor of $Q_r(s)$ and $Q''_e(s)$ a divisor of $Q_p(s)$. The strictly-proper rational function

$$U_r(s) = \frac{P_e(s)}{Q'_e(s)} \qquad (4.13)$$

can be considered the free response, due to a nonzero initial condition, of the part of the internal model contained in the regulator.

The signal generated by Σ_c is described with its $\mathcal{L}$-transform

$$U_c(s) = \frac{P_c(s)}{Q_c(s)} \qquad (4.14)$$

where $P_c(s)$ and $Q_c(s)$ will be determined to obtain perfect tracking of the signal $y_e(t)$ corresponding to (4.10), under the assumption that filter, regulator and plant are in the zero state at $t = 0$. The states of the internal model and compensator can be suddenly changed at any instant of time by the supervising unit. In our framework, referring to $\mathcal{L}$-transforms, the action of the supervising unit can be viewed as delivering Dirac impulses to (4.10), (4.13) and (4.14). If the plant is minimum-phase, perfect tracking can easily be performed without any special artifice, as the following theorem states.

Theorem 4.2. (Perfect tracking in the minimum-phase case). Let us assume that Σ_p is minimum-phase and without any zero on the imaginary axis. The control system of Fig. 6 allows perfect tracking with bounded u_c if and only if the following *relative degree condition* holds:

$$n - m \leq \ell + n_f - 1 \tag{4.15}$$

Proof. If. It can be easily seen that tracking error e can be maintained at zero for a suitable value of n_f and suitable polynomials $P_e(s)$, $P_c(s)$ and $Q_c(s)$. In fact, from

$$E(s) = \frac{\alpha}{s^\ell} \frac{k_f}{Q_f(s)} - \left(\frac{P_e(s)}{Q'_e(s)} + \frac{P_c(s)}{Q_c(s)} \right) \frac{P_p(s)}{Q_p(s)} \tag{4.16}$$

by assuming $Q'_p(s) := Q_p(s)/Q''_e(s)$, we obtain

$$\frac{\alpha \, k_f \, Q'_p(s)}{Q'_e(s) \, Q_f(s) \, P_p(s)} = \frac{P_e(s)}{Q'_e(s)} + \frac{P_c(s)}{Q_c(s)} \tag{4.17}$$

This equality holds if the member on the left is strictly proper, i.e., if (4.15) holds.

Only if. Suppose the control system of Fig. 6 allows perfect tracking, i.e., $y_f(t) = y_p(t)$ $t \in \mathbf{R}$. But, y_f is continuously differentiable at $t = 0$ to the order $\ell + n_f - 2$. Since the input to the plant is piecewise continuous by assumption, $y_f(t)$ is differentiable up to the order of $n - m - 1$. Therefore, clearly $\ell + n_f - 2 \geq n - m - 1$, from which (4.15) follows. □

The proposed solution is not feasible in practice if the plant is non-minimum-phase since the compensator, having the zeros of the plant as poles, is unstable, hence injects unbouded functions at the control input u. On the other hand, we have a remedy for this: preaction. Let us factorize the numerator of $G_p(s)$ as $P_p(s) = P_p^-(s) P_p^+(s)$, with $P_p^-(s)$ and $P_p^+(s)$ having all roots with strictly negative and strictly positive real parts respectively and decompose the last term on the right of (4.17) as

$$\frac{P_c(s)}{Q_c(s)} = \frac{P_{cs}(s)}{P_f(s) \, P_p^-(s)} + \frac{P_{cu}(s)}{P_p^+(s)} \tag{4.18}$$

thus introducing two kinds of open-loop compensators, which are categorized as the *postaction* (meaning after switching) and *preaction* (meaning before switching or anticipatory) compensator. We will allow the preaction to start at a suitable instant of time $t_{pre} < 0$. While the postaction compensator is a linear autonomous system whose state can be suitably changed by the supervising unit, the preaction compensator can be realized as a autonomous unstable linear system with a suitable final state and a "small" in norm initial state: its practical implementation is based on recording the sampled backward solution of the state equation and forcing the compensator to follow this solution in the forward mode by correcting its state at the sampling instants, considered in the reverse order with respect to the backward solution. In Fig. 8 a typical output signal of the compensator is shown, referring to a case where both preaction and postaction are needed. With the above system description and

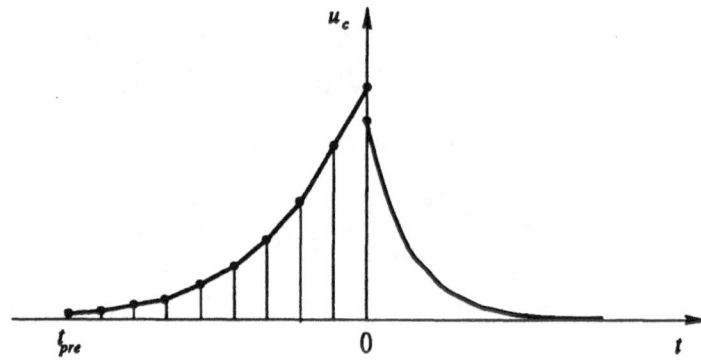

Figure 8: A typical output signal of the compensator, with preaction and postaction.

assumptions we state the main result on SISO almost perfect tracking with preaction in the following terms.

Theorem 4.3. (Almost Perfect Tracking with Preaction). Let us assume that Σ_p is nonminimum-phase, but with no zeros on the imaginary axis. Denote by $t_{pre} < 0$ the time at which the control action starts, i.e., the smallest t such that $u(t) \neq 0$, and assume that the plant state is zero at $t = t_{pre}$. It is possible to realize a control system as shown in Fig. 6 with bounded u_c such that

$$|e(t)| \to 0 \quad \text{as} \quad t_{pre} \to -\infty \quad \text{uniformly in } \mathbf{R}; \qquad (4.19)$$

$$\int_{-\infty}^{+\infty} |e(t)|^2 \, dt \to 0 \quad \text{as} \quad t_{pre} \to -\infty, \qquad (4.20)$$

if and only if the relative degree condition (4.15) holds.

Note that (4.19) and (4.20) imply perfect tracking for an infinite pre-action time. The main result on almost perfect tracking with preaction is stated as follows.

To prove Theorem 4.3 we need the following lemma.

Lemma 4.1. (Time Axis Inversion). Let us consider a continuous-time causal SISO system Σ with transfer function $P(s)/Q(s)$ and denote by $P_0(s)/Q(s)$ the $\mathcal{L}$-transform of a generic free response of Σ. Let $u_r(t)$, $t \in [-\infty, 0]$, be an input function that, starting from the zero state at $t = -\infty$, takes Σ at $t = 0$ to the state corresponding to the given free response, and $y_r(t)$, $t \in [-\infty, 0]$, the corresponding output function; denote by $U_r(s)$ and $Y_r(s)$ their reverse-time $\mathcal{L}$-transforms, i.e., the $\mathcal{L}$-transforms of the functions $u_r(-\tau)$, $y_r(-\tau)$, $\tau \in [0, \infty]$, defined by

$$U_r(s) = \int_0^\infty u_r(-\tau)\, e^{-s\tau}\, d\tau\ , \quad Y_r(s) = \int_0^\infty y_r(-\tau)\, e^{-s\tau}\, d\tau \quad (4.21)$$

The following relation holds:

$$Y_r(s) = \frac{P(-s)}{Q(-s)}\, U_r(s) - \frac{P_0(-s)}{Q(-s)}\ , \quad (4.22)$$

with

$$\frac{P_0(-s)}{Q(-s)} := \frac{P_0(s)}{Q(s)}\bigg|_{s \leftarrow (-s)}, \quad \frac{P(-s)}{Q(-s)} := \frac{P(s)}{Q(s)}\bigg|_{s \leftarrow (-s)}. \quad (4.23)$$

Proof. Let (A, b, c, d) be a state-space representation of $P(s)/Q(s)$ and $x(0) = x_0$ the initial state corresponding to $P_0(s)/Q(s)$. It is well known that

$$\frac{P(s)}{Q(s)} = c\,(sI - A)^{-1}\, b + d, \quad (4.24)$$

$$\frac{P_0(s)}{Q(s)} = c\,(sI - A)^{-1}\, x_0, \quad (4.25)$$

with $(sI-A)^{-1} = \mathrm{adj}(sI-A)/\det(sI-A)$. Functions $u_r(\cdot)$ and $y_r(\cdot)$ satisfy the state space equations

$$-\dot{x}_r(\tau) = A\, x_r(\tau) + b\, u_r(\tau)\ , \quad x_r(0) = x_0 \quad (4.26)$$
$$y_r(\tau) = c\, x_r(\tau) + d\, u_r(\tau) \quad (4.27)$$

The corresponding state function is $x_r(t)$, $t \in [-\infty, 0]$ (or $x_r(-\tau)$, $\tau \in [0, \infty]$), whose reverse-time $\mathcal{L}$-transform, similar to (4.21), is

$$X_r(s) = \int_0^\infty x_r(-\tau)\, e^{-s\tau}\, d\tau \quad (4.28)$$

The $\mathcal{L}$-transform of (4.26, 4.27) is

$$-s\,X_r(s) + x_0 \;=\; AX_r(s) + b\,U_r(s) \tag{4.29}$$

$$Y_r(s) \;=\; cX_r(s) + d\,U_r(s) \tag{4.30}$$

and provides

$$Y_r(s) = (c\,(-sI - A)^{-1}b + d)\,U_r(s) - c\,(-sI - A)^{-1}\,x_0 \tag{4.31}$$

from which (4.22, 4.23) are derived by comparison with (4.24, 4.25). □

Proof of Theorem 4.3. If. Since the control action must be bounded, the postaction signal whose $\mathcal{L}$-transform is the last term on the right of (4.18) cannot be applied. We shall show, however, that the same effect can be obtained with a suitable preaction signal. In plain terms, this is proved as follows: since this signal is a linear combination of modes corresponding to zeros of the plant, an initial state of the plant exists that nulls its effect on the output; hence, the opposite of this initial state has the same effect as the signal. It follows that perfect tracking is still obtainable if we are able to reach this initial state with a bounded preaction. Let us consider the equality

$$\frac{P_p(s)}{Q_p(s)}\frac{P_{cu}(s)}{P_p^+(s)} = \frac{P_p^-(s)\,P_{cu}(s)}{Q_p(s)} \tag{4.32}$$

where the term on the right is the equivalent initial state and denote by $u_{c,pre}(t)$, $t \in [-\infty, 0]$, the preaction signal, that can be derived by using Lemma 4.1. Let $U_{c,pre}(s)$ be the $\mathcal{L}$-transform of the reverse-time function $u_{c,pre}(-\tau)$, $\tau \in [0, \infty]$. From (4.22) that in this case becomes

$$Y_r(s) = \frac{P_p(-s)}{Q_p(-s)}\,U_{c,pre}(s) - \frac{P_p^-(-s)\,P_{cu}(-s)}{Q_p(-s)} \tag{4.33}$$

we obtain that the preaction signal whose reverse-time $\mathcal{L}$-transform is

$$U_{c,pre}(s) = \frac{P_{cu}(-s)}{P_p^+(-s)} \tag{4.34}$$

takes the plant to the desired state while maintaining the output identically zero. Thus error can be nulled if $t_{pre} \to -\infty$. The convergence properties (4.19) and (4.20) will be proved for the MIMO case in Theorem 4.5 of the next subsection.

Only if. This part of the proof is the same as for Theorem 4.2.□

Remarks:

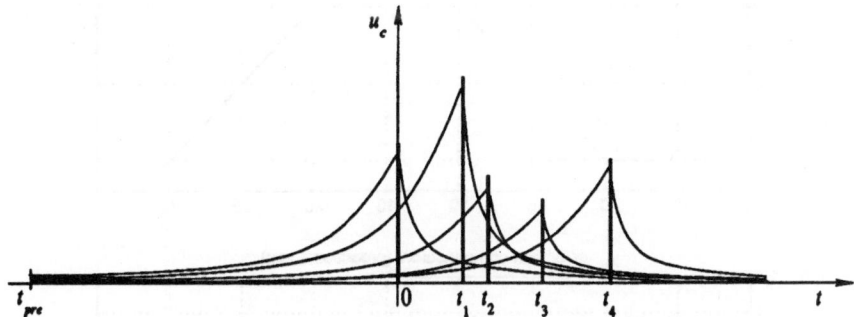

Figure 9: A typical correction signal in the case of infinite preview.

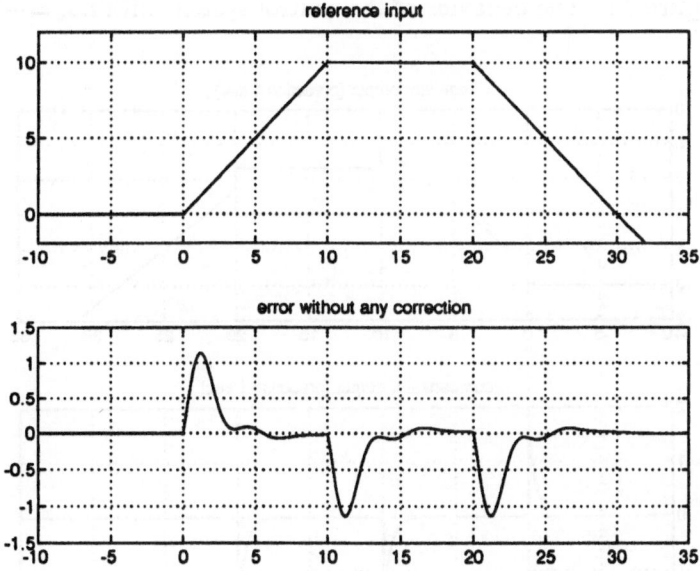

Figure 10: The behavior of the control loop without any correction.

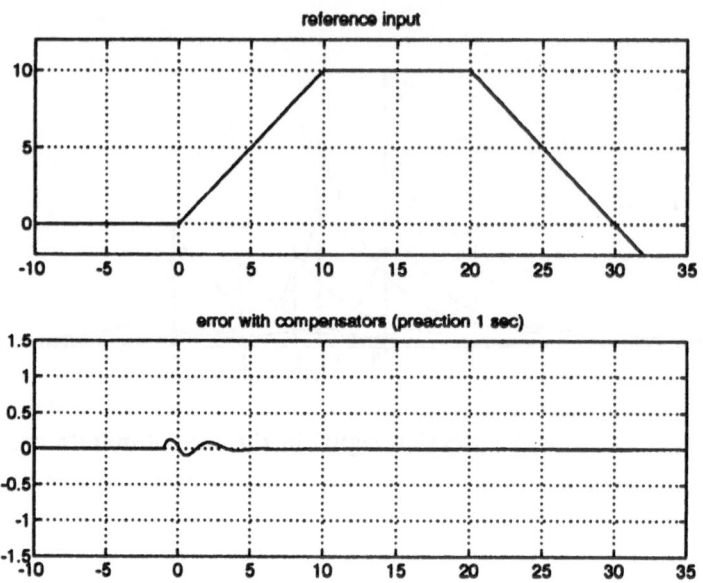

Figure 11: The behavior of the control system with $t_{pre} = -1$.

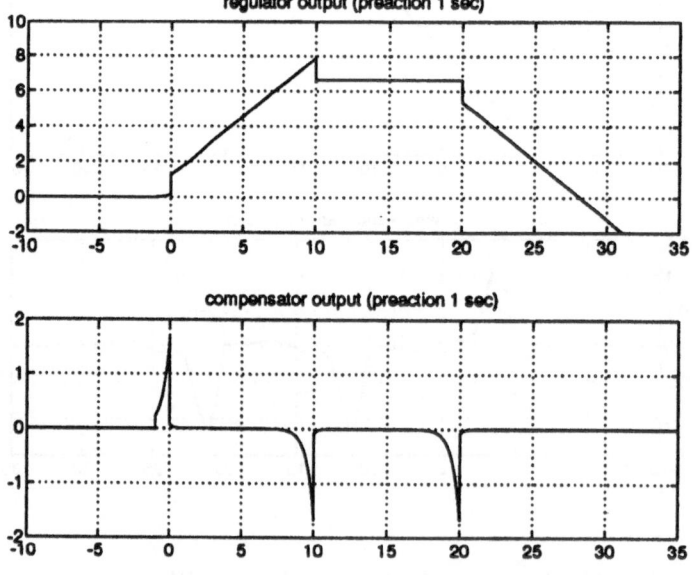

Figure 12: The outputs of the regulator and compensator with $t_{pre} = -1$.

1. A delay $e^{-t_0 s}$ in the plant can be easily handled by the supervising unit by advancing all the correcting signal to the regulator and compensator with respect to the reference.

2. Preaction time is not necessarily equal to preview time. For instance, consider a machine-tool that must reproduce a profile completely known in advance: the preview horizon is infinity, since we can impose any movement after the profile has been followed without error (for instance, a permanent stop condition), while preaction time has to be chosen to neutralize the effects of the unstable zeros of the plant, and thus is related to the maximum time constant corresponding to their real parts. In this case if error is zero at the first exosystem switching, it remains zero during all tracking since in preaction all future switchings have been accounted for. This is clarified by Fig. 9, where the correcting pulses for both preaction and postaction for an infinite preview horizon are shown: since the preaction time increases, perfect tracking is swiftly achieved, after a unique small initial transient.

4.4 A Numerical Example

The above concepts will be illustrated by a simple numerical example. Let

$$H_p(s) = \frac{-3 (s - 2) (s + 4)}{(s + 1) (s + 2) (s + 8)} \qquad (4.35)$$

be the transfer function of the plant, and

$$H_r(s) = \frac{(s + .5)^2 + .3^2}{s^2} \qquad (4.36)$$

that of a standard regulator with a double pole at the origin, required to track a ramp with asymptotic zero error. In this case according to the relative degree condition a filter is not required to follow multiple-switching ramps of the types shown in Fig. 7 Fig. 10 shows the reference input and the error of the standard control loop. Fig. 11 shows the error, the regulator output and the compensator output corresponding to the overall implementation scheme represented in Fig. 6 with $t_{pre} = -1$ sec. The supervising unit in this case drives the unstable zero compensator state along the unstable trajectory starting a t_{pre} (by imposing backward-computed samples, as shown in Fig. 8) and changes the internal model and stable zero compensator states at the switching times.

Note that the error transient is significantly reduced and is present only at the first switching, since an infinite preview horizon has been assumed; if the preview horizon was equal to the preaction time (i.e., switching times were known only one second in advance), the same error transient would be present at every switching. Fig. 12 shows the regulator and compensator output in the same case.

The preaction time $t_{pre} = -5$ sec reduces the error to practically zero (its maximum absolute value during transient after t_{pre} is about 10^{-4}) and maintains it at zero at every switching time.

4.5   Perfect Tracking in MIMO Continuous-Time Systems

The control system layout of Fig. 6 is repeated in Fig. 13 for the MIMO case: it includes a standard multivariable feedback loop consisting of a *plant* Σ_p and a *regulator* Σ_r, designed to react to and asymptotically reject inaccessible and unpredictable disturbances possibly applied to the plant and not shown in the figure. The scheme also includes a *supervising unit* whose aim is to realize, if possible, tracking with zero error of vector reference input $r(t) = (r_1(t), \ldots, r_q(t))$ by suitably killing tracking error transients in the feedback loop.

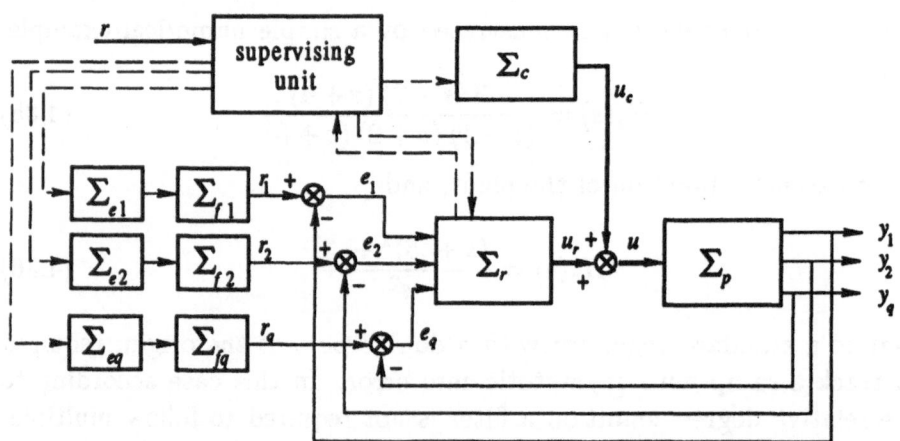

Figure 13: The MIMO perfect or almost-perfect tracking implementation.

To this end the supervising unit: *(i)* generates an approximate replica of the reference input by changing, at discrete instants of time, the leading states of the *exosystems* $\Sigma_{e,i}$, each followed by a *filter* $\Sigma_{f,i}$ $(i = 1, \ldots, q)$, whose purpose is to apply sufficiently smooth signals to the regulation loop; *(ii)* reads and changes the regulator state when necessary; *(iii)* changes the state of a *feedforward compensator* Σ_c when necessary, thus generating correcting signals that are injected on the manipulable input of the plant.

In practice the regulator, exosystems, filters, feedforward compensator and supervising unit can be realized as a single special-purpose digital computer. However, for a neater presentation of the results, in this paper continuous-time models will be adopted.

The plant Σ_p is modelled by the triple (A_p, B_p, C_p) with state dimension n_p, and is assumed to be *functionally controllable*, i.e., according to Property 2.9, satisfying

$$S^* + C = \mathcal{X}_p \tag{4.37}$$

with $S^* := \min S(A_p, B_p, C_p)$, $C := \ker C_p$, and $\mathcal{X}_p := \mathbf{R}^{n_p}$.

If (4.37) holds, by using Algorithm 2.10 it is possible to define a *relative degree* ρ_i, $(i=1,\ldots,q)$, referred to the outputs of the plant.

We assume that the generic i-th exosystem, of order $n_{e,i}$, is represented by the observable pair $(C_{e,i}, A_{e,i})$ $(i=1,\ldots,q)$, with

$$A_{e,i} = \begin{bmatrix} 0 & 1 & \cdots & 0 \\ 0 & 0 & \cdots & 0 \\ \vdots & \vdots & \ddots & \vdots \\ -\eta_{i,0} & -\eta_{i,1} & \cdots & -\eta_{i,n_{e,i}-1} \end{bmatrix}, \quad C_{e,i} = \begin{bmatrix} 1 & 0 & \cdots & 0 \end{bmatrix}$$

$$\tag{4.38}$$

The filters are represented by strictly stable triples $(A_{f,i}, B_{f,i}, C_{f,i})$, of orders $n_{f,i}$ $(i=1,\ldots,q)$. Like the exosystems, the filters are given in the observer canonical form, i.e.,

$$A_{f,i} = \begin{bmatrix} 0 & 1 & \cdots & 0 \\ 0 & 0 & \cdots & 0 \\ \vdots & \vdots & \ddots & \vdots \\ -\varphi_{i,0} & -\varphi_{i,1} & \cdots & -\varphi_{i,n_{f,i}-1} \end{bmatrix}, \quad B_{f,i} = \begin{bmatrix} 0 \\ 0 \\ \vdots \\ \varphi_{i,0} \end{bmatrix},$$

$$C_{f,i} = \begin{bmatrix} 1 & 0 & \cdots & 0 \end{bmatrix} \tag{4.39}$$

that guarantees absence of zeros, hence relative degree $n_{f,i}$.

It may happen that some filters are not necessary in the overall system represented in Fig. 13. To include also this case we represent filters with quadruples $(A_{f,i}, B_{f,i}, C_{f,i}, D_{f,i})$ with the agreement that $D_{f,i}=1$ if the other matrices are empty and $D_{f,i}=0$ otherwise. The overall filtering system can be represented by the quadruple (A_f, B_f, C_f, D_f), where matrices A_f and D_f are block-diagonal built with those of the single filters, while B_f and C_f are built in the same way, but completed with zero columns or rows to save dimensional congruence when dynamics is absent.

For the multivariable loop we use the standard state-space representation whose equations are of the type (3.17–3.19), that model the plant, filters and exosystem as a unique triple (A, B, E)

It is possible to include filters in matrix A of (3.17) by redefining the submatrices (3.18) as

$$x_1 := \begin{bmatrix} x_p \\ x_f \end{bmatrix}, \quad A_1 := \begin{bmatrix} A_p & O \\ O & A_f \end{bmatrix}, \quad A_3 := \begin{bmatrix} O \\ B_f C_e \end{bmatrix}, \quad B_1 := \begin{bmatrix} B_p \\ O \end{bmatrix} \quad (4.40)$$

$$x_2 := x_e, \quad E_1 := [-C_p \quad C_f], \quad E_2 := D_f C_e \quad (4.41)$$

so that $x_1 \in \mathbf{R}^{n_1}$, with $n_1 = n_p + n_f$, now denotes the state of the system "plant plus filters". Note that (A_1, B_1) is still stabilizable, (A, E) detectable, and (3.38), (3.39) still hold. Furthermore, the regulator designed for the plant in the absence of filters still satisfies both the asymptotic regulation condition and the loop stability condition, since the modes of the filters tend to zero as time approaches infinity, like those of the control loop.

The feedforward compensator is represented by the observable pair (C_c, A_c), of order n_c. We state the following theorem, that extends Theorem 4.2 to the MIMO case.

Theorem 4.4. (Perfect tracking in the minimum-phase case). Assume that *(i)* Σ_p is functionally controllable, *(ii)* Σ_p is minimum-phase (i.e., $\mathcal{Z}(A_p, B_p, C_p) \subseteq \mathbf{C}_-$). The control system of Fig. 13 allows perfect tracking with bounded u_c if and only if the following *relative degree condition* holds:

$$n_{f,i} \geq \rho_i - n_{e,i} + 1 \quad (i = 1, \ldots, q) \quad (4.42)$$

Proof. If. Let $x_{e,i}(0+) = (0, 0, \ldots, \alpha_i)^T$ $(i = 1, \ldots, q)$ be the column vectors of the initial states of the exosystems and denote with $x_2(0+)$ the overall exosystem initial state, a column vector having these vectors as elements. Consider the basis matrix (3.27) for $\hat{\mathcal{W}}$ referring to the regulator problem with matrices defined by (4.40, 4.41), i.e., with filters modelled as an uncontrollable part of the plant. The overall system initial state corresponding to tracking without transients is computed as

$$\begin{bmatrix} x_1(0+) \\ x_2(0+) \\ z(0+) \end{bmatrix} = \begin{bmatrix} X_1 \\ I_{n_2} \\ Z \end{bmatrix} x_2(0+). \tag{4.43}$$

tracking by means of a suitable control function is possible. This implies that state $x_1(0+)$ belongs to an $(A_1, \mathrm{im}B_1)$-controlled invariant contained in $\ker E_1$, hence to $\mathcal{V}_1^* = \max\mathcal{V}(A_1, \mathrm{im}B_1, \ker E_1)$, since only if this is so does a corrective control function $u_c(t)$ exist that, starting from $-x_1(0+)$, maintains the plant state trajectory on $\mathcal{V}_1^*$, hence with no influence on the error variable. By superposition, this control function keeps the error variable at zero if applied with our actual initial state $(0, x_2(0+), z(0+))^T$. It is easily shown that $u_c(t)$ can be generated by the compensator: let us denote by n_c the dimension of $\mathcal{V}_1^*$ and by T_1 a $n_1 \times n_c$ basis matrix of $\mathcal{V}_1^*$. Let F_1 be a matrix such that $(A_1 + B_1 F_1)\mathcal{V}_1^* \subseteq \mathcal{V}_1^*$, define $T = [T_1\ T_2]$ with T_2 such that T is nonsingular and assume as A_c the first $n_c \times n_c$ submatrix of $T^{-1}(A_1 + B_1 F_1)T$ and as C_c the $p \times n_c$ matrix $F_1 T_1$. The initial state $x_c(0+)$ of the compensator is then given by the first n_c components of $-T^{-1} x_1(0+)$. The internal unassignable eigenvalues of $\mathcal{V}_1^*$ are the union of the elements of $\mathcal{Z}(A_p, B_p, C_p)$ (the invariant zeros of the plant) and of the eigenvalues of the filters, that appear as invariant zeros of (A_1, B_1, E_1) since filters are not controllable. Hence the corresponding eigenstructures are repeated in A_c.

Only if. We suppose that the supervising unit reproduces the reference signal by changing the first state variable of the integrator chain present in every exosystem at discrete instants of time, so that the ouput of the i-th exosystem has the $(n_{e,i} - 1)$-th time derivative piecewise continuous, while the i-th output can reproduce any function with ρ_i-th derivative piecewise continuous, so that a filter satysfying (4.42) is necessary to restrict the class of reference functions.   □

Remark: The role of the compensator is to drive the state of the plant and filters along trajectories on the $(A_1 + B_1 F_1)$-invariant $\mathcal{V}_1^*$, thus making it possible to reproduce in the controlled plant the effect of state feedback matrix F_1 for every initial state belonging to $\mathcal{V}_1^*$. Owing to Theorem 4.4 it is possible to obtain an asymptotically stable compensator if and only if the plant is minimum phase. Hence condition (15) is sufficient to ensure zero tracking error of the replicated reference signal with bounded input. Note that F_1 can and must also stabilize the internal assignable eigenvalues of $\mathcal{V}_1^*$ if the controlled plant is not invertible.

Theorem 4.5. (Almost Perfect Tracking with Preaction). Assume that *(i)* Σ_p is functionally controllable, *(ii)* $\mathcal{Z}(A_p, B_p, C_p) \cap C_0 = \emptyset$. Denote by $t_{pre} < 0$ the time at which the control action starts, i.e., the smallest t such that $u(t) \neq 0$, and assume that the plant state is zero at $t = t_{pre}$. It is possible to realize a control system as shown in Fig. 13 with bounded u_c such that

$$\|e(t)\| \to 0 \quad \text{as } t_{pre} \to -\infty \quad \text{uniformly in } \mathbf{R}; \tag{4.44}$$

$$\int_{-\infty}^{+\infty} \|e(t)\|^2 \, dt \to 0 \quad \text{as } t_{pre} \to -\infty, \tag{4.45}$$

if and only if the relative degree condition (4.42) holds.

   Proof. If. Since $\mathcal{Z}(A_p, B_p, C_p)$ has no purely imaginary elements, of A_c can be partitioned into a strictly stable set and a strictly unstable set. A similarity transformation T_c exists such that

$$T_c^{-1} A_c T_c = \begin{bmatrix} A_{c,s} & O \\ O & A_{c,u} \end{bmatrix} \tag{4.46}$$

where $A_{c,s}$ has the strictly stable eigenvalues and $A_{c,u}$ those strictly unstable. Without loss of generality, we can assume that transformation T_c is included in T_1, so that A_c has the structure shown on the right of (4.46) and the initial state $x_c(0+)$ of the compensator is accordingly partitioned as $x_{c,s}(0+)$, $x_{c,u}(0+)$. Let

$$x_{1,s}(0+) := T_1 \begin{bmatrix} x_{c,s}(0+) \\ 0 \end{bmatrix}, \qquad x_{1,u}(0+) := T_1 \begin{bmatrix} 0 \\ x_{c,u}(0+) \end{bmatrix} \tag{4.47}$$

as far as $x_{c,s}(0+)$ is concerned, we proceed as before, generating a control action $u_2(t)$ corresponding to an output-invisible state trajectory $x_1(t)$ of the plant and filters starting from $-x_{1,s}(0+)$ and tending to zero as time approaches infinity, while for $x_{c,u}(0+)$ we integrate the compensator equation backward in time, i.e., with respect to the time variable $\tau := -t$, thus generating again a stable trajectory $x_1(\tau)$ and a stable control function $u_2(\tau)$, tending to zero as τ approaches infinity. Since the reverse-time trajectory $x_1(t) = x_1(-\tau)$, $t \in [-\infty, 0]$ is admissible and externally stable because of the strict stability of the control loop, by applying the control function $u_2(t) = u_2(-\tau)$, $t \in [-\infty, 0]$ we obtain a zero-output trajectory that reaches $x_{1,u}(0+)$ at $t = 0$.

   Let us denote by $t_{pre} < 0$ the time at which the control action starts (i.e., the smallest t such that $u(t) \neq 0$), and assume that the overall system state is zero at $t = t_{pre}$. Tracking error would be zero if the state of the plant (including filters) at $t = t_{pre}$ were

$$x_1(t_{pre}) = T_1 \, e^{A_{c,u} t_{pre}} \begin{bmatrix} 0 \\ x_{c,u}(0+) \end{bmatrix} \qquad (4.48)$$

but, since the control action starts at t_{pre}, it is zero. The error transient can be computed by considering the response to $-x_1(t_{pre})$ of the extended plant (that, in this case, consists of plant, filters and regulator) with the regulator state $z(t_{pre})$ at zero. Let us denote with $\alpha = (\alpha_1, \ldots \alpha_q)$ the vector of the exosystems initial states and let σ_1 (> 0) be the minimal real part of the eigenvalues of $A_{c,u}$; since $x_{c,u}(0+)$ is a linear function of α, a constant $\delta_1 > 0$ exists such that

$$\|x_1(t_{pre})\| \leq \delta_1 \, e^{\sigma_1 t_{pre}} \|\alpha\| \qquad (4.49)$$

Let

$$A_{ep} := \begin{bmatrix} A_1 + B_1 K E_1 & B_1 L \\ M E_1 & N \end{bmatrix}, \quad E_{ep} := \begin{bmatrix} E_1 & O \end{bmatrix} \qquad (4.50)$$

be the matrices of the extended plant. The error is given by

$$e(t - t_{pre}) = E_{ep} \, e^{A_{ep}(t - t_{pre})} \begin{bmatrix} -x_1(t_{pre}) \\ 0 \end{bmatrix}, \quad t \geq t_{pre} \qquad (4.51)$$

Let us denote by σ_2 (< 0) the maximal real part of the eigenvalues of A_{ep}: a constant $\delta_2 > 0$ exists such that

$$\|e(t)\| \leq \delta_2 \, e^{\sigma_2(t - t_{pre})} \|x_1(t_{pre})\| \leq \delta_1 \, \delta_2 \, e^{\sigma_2(t - t_{pre})} e^{\sigma_1 t_{pre}} \|\alpha\|$$
$$\leq \delta_1 \, \delta_2 \, e^{\sigma_1 t_{pre}} \|\alpha\| \qquad (4.52)$$

Furthermore,

$$\int_{-\infty}^{\infty} \|e(t)\|^2 \, dt = \int_{t_{pre}}^{\infty} \|e(t)\|^2 \, dt \leq \int_{t_{pre}}^{\infty} \delta_1^2 \, \delta_2^2 \, e^{2\sigma_1(t - t_{pre})} e^{2\sigma_1 t_{pre}} \|\alpha\| \, dt$$

$$= \frac{\delta_1^2 \, \delta_2^2}{-2\sigma_2} e^{2\sigma_1 t_{pre}} \|\alpha\| . \qquad (4.53)$$

The statement directly follows from (4.52) and (4.53).

Only if. This part of the proof is as for Theorem 4.4. □

Corollary 4.1. Let $\alpha = (\alpha_1, \ldots \alpha_q)$ be the vector of the exosystems initial states. Then, given any number $\epsilon > 0$, we have the following:

$$t_{pre} \leq \frac{1}{\sigma_1} \ln \frac{\epsilon}{\delta_1 \delta_2 \|\alpha\|} \quad \text{implies} \quad \|e(t)\| \leq \epsilon, \quad t \in \mathbf{R} \qquad (4.54)$$

$$t_{pre} \leq \frac{1}{2\sigma_1} \ln \frac{-2\epsilon\sigma_2}{\delta_1^2 \delta_2^2 \|\alpha\|} \quad \text{implies} \quad \int_{-\infty}^{+\infty} \|e(t)\|^2 \, dt \leq \epsilon \qquad (4.55)$$

where the quantities t_{pre}, σ_1, σ_2, δ_1, δ_2 and $e(t)$ are defined as in the proof of Theorem 4.5.

Remarks:

1. The proofs of Theorems 4.4 and 4.5 suggest a working algorithm for the supervising unit to deal with exosystems state switching at $t = 0$. Since the overall system is linear and time-invariant, multiple switchings at different instants of time are easily handled by superposition and time-shifting. Note that the regulator state corresponding to tracking without error must be superimposed on the current regulator state, that may be affected by disturbances to be neutralized; hence at every switching the supervising unit must read the current regulator state and apply to it the increment computed with the above algorithm. The same applies to the compensator, whose state, subject to multiple transients, is incremented at every switching by the amount derived by means of the algorithm.

2. The proposed control scheme realizes perfect or almost-perfect tracking control in multivariable systems by injecting suitable correcting pulses at the control input: of course, perfect tracking implies noninteraction, hence it is shown that noninteraction is a second-degree-of-freedom property.

Many standard problems of multivariable control theory for which a solution was already given by using geometric techniques, can be revisited with the new concept of preaction, that allows trading causality for stabilizability. An example is given in the following exercise.

Exercise 4.1. Let us refer to the disturbance localization problem (Section 2.6) and assume that the disturbance $d(\cdot)$ is accessible and completely known a priori. Show that the accessible disturbance localization problem with stability has a solution if and only if (A, B) is stabilizable and (2.43) holds.

Notes and References. The model-reference two-degrees-of-freedom method to deal with tracking problems, both in continuous and in discrete-time cases, is well-knowm. It is also widely used in adaptive control systems (see for instance Åström and Wittenmark [1, 2] and Mosca [45]). It is also known that in LQ optimal tracking and in model-predictive control, preview gives significant benefits in the nonminimum-phase case. Important results in discrete-time control systems using preaction by special design of the feedforward path of a two-degrees-of-freedom controller were obtained by Jayasuriya and Tomizuka [33] and Gross, Tomizuka and Messner [29]. The use of a two-input controller with a special digital unit to deal with the infinite or time-varying preview case in the discrete-time case has been proposed by Marro and Fantoni [37]. In the continuous-time case basic results on system inversion and control with preaction, based on the backward computation of the unstable zeros dynamics were presented by Devasia, Paden and Rossi [22, 23] and Hunt, Meyer and Su [31]. The use of a supervising unit feeding compensators that generate correcting pulses at discrete instants of time was applied by Marro and Piazzi to eliminate regulation transients in the presence of large parameter jumps or structural changes in minimum-phase multivariable controlled plants [39].

5 References

[1] Åström, K.J., and B. Wittenmark, *Computer-Controlled Systems: Theory and Design* Prentice Hall, Englewood Cliffs, N.J., 1990.

[2] Åström, K.J., and B. Wittenmark, *Adaptive Control, (2nd edition)* Addison-Wesley, Reading, Massachusetts, 1995.

[3] Basile, G., Laschi, R. and G. Marro, "Invarianza controllata e non interazione nello spazio degli stati," *L'Elettrotecnica*, vol. LVI, n. 1, 1969.

[4] Basile, G., and G. Marro, "Controlled and conditioned invariant subspaces in linear system theory," *J. Optimiz. Th. Applic.*, vol. 3, no. 5, pp. 305–315, 1969.

[5] Basile, G., and G. Marro, "On the observability of linear time-invariant systems with unknown inputs," *J. of Optimiz. Th. Applic.*, vol. 3, no. 6, pp. 410–415, 1969.

[6] Basile, G., and G. Marro, "L'invarianza rispetto ai disturbi studiata nello spazio degli stati," *Rendiconti della LXX Riunione Annuale AEI*, paper 1-4-01, Rimini, Italy, 1969.

[7] Basile, G., and G. Marro, "A state space approach to noninteracting controls," *Ricerche di Automatica*, vol. 1, no. 1, pp. 68–77, 1970.

[8] Basile, G., and G. Marro, "On the perfect output controllability of linear dynamic systems," *Ricerche di Automatica*, vol. 2, no. 1, pp. 1–10, 1971.

[9] Basile, G., and G. Marro, "Self-bounded controlled invariant subspaces: a straightforward approach to constrained controllability," *J. Optimiz. Th. Applic.*, vol. 38, no. 1, pp. 71–81, 1982.

[10] Basile, G., and G. Marro, "On the robust controlled invariant," *Systems & Control Letters*, vol. 9, no. 3, pp. 191–195, 1987.

[11] Basile, G. and G. Marro, *Controlled and Conditioned Invariants in Linear System Theory*, Prentice Hall, New Jersey, 1992.

[12] Basile, G., G. Marro, and A. Piazzi, "A new solution to the disturbance localization problem with stability and its dual," *Proceedings of the '84 International AMSE Conference on Modelling and Simulation*, vol. 1.2, pp. 19–27, Athens, 1984.

[13] Basile, G., G. Marro, and A. Piazzi, "Stability without eigenspaces in the geometric approach: some new results," *Frequency Domain and State Space Methods for Linear Systems*, edited by C. A. Byrnes and A. Lindquist, North-Holland (Elsevier), Amsterdam, pp. 441–450, 1986.

[14] Cevik, M.K.K. and J.M. Schumacher, "The regulator problem with robust stability," *Automatica*, vol. 31, pp. 1393–1406, 1995.

[15] Chen, C.T., *Linear system theory and design*, Holt, Rinehart and Winston, New York, 1984.

[16] Davison, E.J., "The robust control of a servomechanism problem for linear time-invariant multivariable systems," *IEEE Trans. on Autom. Contr.*, no. 21, pp. 25–33, 1976.

[17] Davison, E.J., and I.J. Ferguson, "The design of controllers for the multivariable robust servomechanism problem using parameter optimization methods," *IEEE Trans. on Autom. Contr.*, vol. AC-26, no. 1, pp. 93–110, 1981.

[18] Davison, E.J., and A. Goldemberg, "Robust control of a general servomechanism problem: the servo compensator," *Automatica*, vol. 11, pp. 461–471, 1975.

[19] Davison, E.J., and B.M. Scherzinger, "Perfect control of the robust servomechanism problem," *IEEE Trans. on Autom. Contr.*, vol. AC-32, no. 8, pp. 689–701, 1987.

[20] Davison, E.J., and S.H. Wang, "Properties and calculation of transmission zeros of linear multivariable systems," *Automatica*, vol. 10, pp. 643–658, 1974.

[21] Davison, E.J., and S.H. Wang, "Remark on multiple transmission zeros" (correspondence item), *Automatica*, vol. 10, pp. 643–658, 1974.

[22] Devasia, S. and B. Paden, "Exact output tracking for nonlinear time-varying systems", *Proc. IEEE Conf. on Decision and Control*, Lake Buena Vista, CA, pp. 2346–2355, 1994.

[23] Devasia, S., B. Paden, and C. Rossi, "Minimal transient regulation of flexible structures." *Proceedings of the IECON '94*, Bologna, Italy, Sept. 5-9 1994.

[24] Francis, B.A., "The linear multivariable regulator problem," *SIAM J. Contr. Optimiz.*, vol. 15, no. 3, pp. 486–505, 1977.

[25] Francis, B.A., "The multivariable servomechanism problem from the input-output viewpoint," *IEEE Trans. Autom. Contr.*, vol. AC-22, no. 3, pp. 322–328, 1977.

[26] Francis, B., O.A. Sebakhy, and W.M. Wonham, "Synthesis of multivariable regulators: the internal model principle," *Applied Math. & Optimiz.*, vol. 1, no. 1, pp. 64–86, 1974.

[27] Francis, B.A., and W.M. Wonham, "The role of transmission zeros in linear multivariable regulators," *Int. J. Control*, vol. 22, no. 5, pp. 657–681, 1975.

[28] Francis, B.A., and W.M. Wonham, "The internal model principle of control theory," *Automatica*, no. 12, pp. 457–465, 1976.

[29] Gross, E., M. Tomizuka and W. Messner, "Cancellation of discrete time unstable zeros by feedforward control," *ASME J. Dynamic Syst., Measurement Contr*, vol. 116, no. 1, pp. 33–38, Mar. 1994.

[30] Hamano, F. and G. Marro, "Using preaction to eliminate tracking error in feedback control of multivariable systems," *DEIS Report no. GA-3-94*, University of Bologna, Italy, 1994 (also submitted to CDC'96, Osaka, Japan).

[31] Hunt, L.R., G. Meyer and R. Su, "Output tracking for nonlinear systems," *Symposium on Implicit and Nonlinear Systems*, Arlington, Texas, pp. 314–320, 1992.

[32] Isidori, A., *Nonlinear Control System (3rd edition)*, Springer-Verlag, Berlin, 1995.

[33] Jayasuriya, S. and M. Tomizuka, "Feedforward controllers for perfect tracking that does not invert plant zeros outside the unit disk," *Proc. 12th IFAC World Congress*, vol. 3, pp. 91–94, 1993.

[34] Kučera, V., *Discrete linear control*, John Wiley & Sons, New York, 1979.

[35] Laschi, R., and G. Marro, "Alcune considerazioni sull'osservabilità dei sistemi dinamici con ingressi inaccessibili," *Rendiconti della LXX Riunione Annuale AEI*, paper 1-4-01, Rimini, Italy, 1969.

[36] Marro, G., "Controlled and conditioned invariants in the synthesis of unknown-input observers and inverse systems," *Control and Cybernetics* (Poland), vol. 2, no. 3/4, pp. 81–98, 1973.

[37] Marro, G. and M. Fantoni, "Using preaction with infinite or finite preview for perfect or almost perfect digital tracking," in press, *Proc. MELECON'96*, Bari, Italy, 1996.

[38] Marro, G. and A. Piazzi (1992), "Feedback systems stabilizability in terms of invariant zeros." In A. Isidori and T. J. Tarn (Ed.), *Systems. Models and Feedback: Theory and Applications*. Birkhäuser series "Progress in Systems and Control Theory", Birkhäuser, Boston.

[39] Marro, G. and A. Piazzi, "Regulation without transients under large parameter jumps." *Proc. 12th IFAC World Congress*, vol. 4, pp. 23–26, 1993.

[40] Marro, G. and A. Piazzi, "A geometric approach to multivariable perfect tracking," in press, *Proc. 13th IFAC World Congress*, San Francisco, 1996.

[41] Middleton, R.H., and G.C. Goodwin, *Digital control and estimation*, Prentice Hall, Englewood Cliffs, N.J., 1990.

[42] Morse, A.S., "Output controllability and system synthesis," *SIAM J. Control*, vol. 9, pp. 143–148, 1971.

[43] Morse, A.S., "Structural invariants of linear multivariable systems," *SIAM J. Control*, vol. 11, pp. 446–465, 1973.

[44] Morse, A.S. and W.M. Wonham, "Status of noninteracting control," *IEEE Trans. Autom. Contr.*, vol. AC-16, no. 6, pp. 568–581, 1971.

[45] Mosca, E, *Optimal, Predictive, and Adaptive Control*, Prentice Hall, Englewood Cliffs, N.J., 1995.

[46] Piazzi, A., and G. Marro, "The role of invariant zeros in multivariable system stability," *Proceedings of the 1991 European Control Conference*, Grenoble, 1991.

[47] Schumacher, J.M.H., "Regulator synthesis using (C, A, B)-pairs," *IEEE Trans. Autom. Contr.*, vol. AC-27, no. 6, pp. 1211–1221, 1982.

[48] Schumacher, J.M.H., "The algebraic regulator problem from the state-space point of view," *Linear Algebra and its Applications*, vol. 50, pp. 487–520, 1983.

[49] Schumacher, J.M.H., "Almost stabilizability subspaces and high gain feedback," *IEEE Trans. Autom. Contr.*, vol. AC-29, pp. 620–627, 1984.

[50] Weiland, S. and J.C. Willems, "Almost disturbance decoupling with internal stability," *IEEE Trans. Autom. Contr.*, vol. 34, pp. 277–286, 1989.

[51] Willems, J.C., "Almost invariant subspaces: an approach to high gain feedback design - Part I: Almost controlled invariant subspaces," *IEEE Trans. Autom. Contr.*, vol. AC-26, no. 1, pp. 235–252, 1981.

[52] Willems, J.C., "Almost invariant subspaces: an approach to high gain feedback design - Part II: Almost conditionally invariant subspaces," *IEEE Trans. Autom. Contr.*, vol. AC-27, no. 5, pp. 1071–1085, 1982.

[53] Willems, J.C., and C. Commault, "Disturbance decoupling by measurement feedback with stability or pole placement," *SIAM J. Contr. Optimiz.*, vol. 19, no. 4, pp. 490–504, 1981.

[54] Wonham, W.M., "Tracking and regulation in linear multivariable systems," *SIAM J. Control*, vol. 11, no. 3, pp. 424–437, 1973.

[55] Wonham, W.M., *Linear Multivariable Control: A Geometric Approach*, Springer-Verlag, New York, 1985.

[56] Wonham, W.M., "Geometric state-space theory in linear multivariable control: a status report," *Automatica*, vol. 15, pp. 5–13, 1979.

[57] Wonham, W.M. and A.S. Morse, "Decoupling and pole assignment in linear multivariable systems: a geometric approach," *SIAM J. Control*, vol. 8, no. 1, pp. 1–18, 1970.

[58] Wonham, W.M., and J.B. Pearson, "Regulation and internal stabilization in linear multivariable systems," *SIAM J. Control*, vol. 12, no. 1, pp. 5–18, 1974.

5. Practice and Trends in Control Engineering

Claudio Maffezzoni*   Gianantonio Magnani*

Abstract

With reference to automatic control, the lecture will analyse the current state of engineering methods available for industrial applications, the principal needs demanding for more systematic and reliable design methods and the lines along which engineering reserch moves to meet such demands. Two different aspects of innovation are considered: the development of new products enhancing the capability of Control Systems (either in stand-alone form or as Distributed Control Systems) to implement and manage more advanced and effective control strategies; the development of tools and frameworks able to improve the work process of engineering projects. Attention is focused on key points for engineering practice: role of standards, data sharing and integration, CAE for control, simulators, and other means suitable to build an actual bridge between Control Science and Engineering. Recent proposals are finally discussed that widen the scope of automatic control to different real-time functions, acting as support for the operator's decision making.

1   Introduction

The role of Engineering as a bridge between science and human works expanded in the last few decades from traditional fields dealing with machines and materials to the rapidly growing area of Information Technol-

*Politecnico di Milano, Dipartimento di Elettronica e Informazione, Piazza Leonardo da Vinci 32, 20133 Milano, Italia.

ogy. Because of the tremendous rate at which Information Technology grows, there is an evident difficulty of the engineering research to develop adequate methods for engineering practice.

In the field of Automatic Control the research community experiences almost simultaneous progresses of science, technology and engineering; as a consequence, it is a very hard task to timely transfer the new findings of science and technology into effective, well proved methods for engineering practice.

Control engineering is concerned with two principal types of activities, depending on what is the scope of the company involved:

1. developing new products, possibly transferring on the market new findings of control science and/or information technology (this is the typical scope of good system vendors);

2. designing and installing systems dedicated to plant-wide control and automation (this is the typical activity of engineering companies).

Big users (for instance large utilities) may have substantial control engineering capacity at home, especially when they are responsible for the basic operating concepts of their plants; in those case, the user itself plays the role of engineering company for the most general aspects.

Whatever is the object of the activity, complexity and sophistication are common features of today control engineering, so that the critical points are:

- limiting and predicting project lead-time;

- managing the interdependent work of many engineers;

- predicting system performance and costs.

As a consequence control engineering demands mainly regard:

- improving procedures and methodologies for system specification, design, documentation, testing and commissioning;

- fixing better standards for control device interoperability and interface;

- maximizing reuse of engineering work to reduce lead-time, cost and uncertainties.

Moreover, when the scope is designing a control system dedicated to a plant or a machine, a key issue is to conceive the control as an integral part of the whole realization because plant operability and performance can not be evaluated without considering control.

With respect to the above problems, the qualification of the procedures and tools adopted by a certain organization to produce "control engineering" plays the same crucial role as in software engineering.

Any means to better define, measure, control and manage the process involved in control engineering project life cycle has to be seriously considered. Among them, the capacity of monitoring and applying technology changes systematically is very important for product innovation.

Moreover, it is recognized [1] that the most profound trend in the engineering contractor environment is in the way projects are executed and, in particular, the drive to utilize integrated date base technology for the execution of engineering design projects.

Thus, three are the aspects that are crucial in control engineering progress: the available technologies to be used, the procedures adopted by the organization to develop projects and the methodologies and tools that help engineers in making procedures effective.

This paper separately analyses the present state and the most interesting trends in control products innovations, on one side, and the actual bottlenecks in control engineering professional work, on the other side, considering both the evolution of procedures and of the CAD-CAE systems most likely involved in that work.

It is then considered the specific role of control science, very often concerned with control algorithms only, in control engineering progress, looking at research lines that are more promising for short and mid term development of new products and of new ways to execute projects.

2　Achievements and trends in product innovation

2.1　Control devices and systems: Current features

Commercial microprocessor and computer based devices employed in modern process control and automation systems may be classified into (stand-alone) industrial controllers, also referred to as PID controllers or single-station controllers, Distributed Control Systems (DCS), and Programmable Logic Controllers (PLC). The main features of these classes of products are outlined in the following, focusing expecially with those more related to control science.

PID controllers

PID controllers are broadly used in small and medium size automation systems of chemical, metallurgical, food and beverage, pharmaceutical, plastics processing, energy, and many other industrial processes. They are also used at subsystem level in large plants, as, for instance, electric power plants, oil and gas production and distribution, and chemical plants. Basically, PID are single-loop controllers and most frequently they are used in temperature, pressure, flow, level and composition control loops. Perhaps temperature control is the most diffuse application. According to a recent study [2], the production of PID temperature controllers is forecast to be about 900 million US$ worldwide in 1997. More than one million controllers are sold each year. Frequently, the effectiveness of PID control is greatly improved by organizing PID according to control paradigms that take advantage of additional measurements and of additional knowledge of the process behaviour. Well known example of such paradigms are cascade, feedforward and ratio control.

Packaging. PID controllers are panel mounting devices. Perhaps for this reason, their size is identified by their front panel dimensions, that commonly adhere to the DIN 43700 standard. The most common sizes are referred to as 1/2DIN (72 × 144mm), 1/4DIN (96 × 96mm), 1/8DIN(48 × 96mm), 1/16DIN (48 × 48mm). The size is a good indicator of the capabilities, functions and cost of a controller. Nevertheless, thanks to advances in microelectronics and surface-mounted component technologies, the new products are offering more capabilities with the same sizes. The most popular size is the 1/8DIN. Together with the 1/16 DIN, they are suitable to be used locally on the (sub)system where the control loop has to be realized. On the contrary, higher class models are usually designed to be installed in a control room. Latest 72 × 144mm and 1/4DIN products are multi-loop controllers, capable of controlling two (typical) and also four or more process variables [3]. Furthermore they are programmable, i.e., the user may build its own control algorithm by composing standard processing and control blocks [4]. Tools for graphic programming running on a PC are commonly adopted to this purpose. Control paradigms like cascade control and the Smith's predictor are usually predefined. Manufacturers proudly refer to such controllers as mini DCS, also to remark their powerful communication capabilities.

Control algorithms and autotuning. The controller's algorithms are not standardized. Regardless of their sizes, almost all commercial controllers implement the basic proportional (P), integral (I) and derivative (D) control modes, but they may differ in the structure of the control law and in several implementation solutions. According to [5], three basic structures are used in the commercial controllers: the standard form, or ISA form, the series form, and the parallel form. Introducing $U(s)$, $Y(s)$, and $Y_{sp}(s)$ as the Laplace transform of process input u, process output y, and setpoint y_{sp}, and $E(s) = Y_{sp}(s) - Y(s)$, the standard form is given by:

$$U = K \left[bY_{sp}(s) - Y + \frac{1}{sT_i}E + \frac{sT_d}{1 + sT_d/N}(cY_{sp}(s) - Y) \right]$$

where K is the controller gain, T_i is the integral time, T_d is the derivative time, N determines the bandwidth of the filter on the derivative action, or limits the derivative gain. Typically $8 \leq N \leq 20$. b and c are weightings that influence the setpoint response, and are usually either 0 or 1 in commercial controllers, but some of these exploit b, $0 \leq b \leq 1$, to improve the setpoint response. In this way the PID behaves as a two-way (or two degrees-of-freedom) controller. The series form is given by:

$$U = K' \left[\left(b + \frac{1}{sT_i'} \right) \frac{scT_d'}{1 + sT_d'/N}Y_{sp} - \left(1 + \frac{1}{sT_i'} \right) \frac{sT_d'}{1 + sT_d'/N}Y \right]$$

and the parallel form by:

$$U = K'' \left[bY_{sp} - Y + \frac{K_i''}{s}E + \frac{sK_d''}{1 + sK_d''/(NK_d'')}(cY_{sp} - Y) \right].$$

Most of higher quality recent products, even in the smallest 1/16 DIN size, are equipped with autotuning capabilities, i.e., they tune their parameters automatically at start-up (some products) and on demand from the user. This greatly simplifies the use of the controllers and reduces the commissioning time. Some products can also adapt continuously their parameters. Even if in the last decade autotuning and continuous adaptation methods have been strongly improved, there are still large margins of progress and matter for research activity. A wide variety of autotuning and continuous adaptation methods are used in commercial products. A look to methods used by several companies that accepted to disclose information about them is reported in [6]. An in depth presentation of most important methods is given in the book of Åström and Hagglund [5], that presents also details on specific implementations in some commercial products. A survey on automatic tuning and adaptation methods for PID controllers is given in [5]. At last, the interested reader may refer to manufacturer's documentation for further details. Even if, in practice a continuous adaptation technique can be used for automatic tuning, many of the adopted autotuning methods are conceived just for one-shot tuning. Autotuning methods consist of two basic steps, namely process dynamics identification and controller parameter design, and each step can be afforded following several diverse ways. Thus methods can be classified according to the adopted identification technique and parameter design formula. As illustrative examples, the major features of two widely used methods are briefly recalled.

Ziegler-Nichols step response method. The first method is the Ziegler and Nichols step response method. It is a time domain, open loop method. A step is applied to the control variable and the process response is analysed automatically to determine the gain, delay, and dominant time constant of a first order with delay model of the process:

$$G(s) = \frac{K_p}{1 + sT} e^{-sL}.$$

Once the parameters of this model have been identified a variety of formulas allow one to calculate the PID parameters. The "right" formula depends on one side on the controllability ratio, also called normalized dead time, defined as:

$$\tau = \frac{L}{L + T},$$

and, on the other side, on the selected performance index (i.e., IAE, ISE, ITAE). Basic formulas are those originally proposed by Ziegler and Nichols [7]. Refined formulas, and discussion of the conditions of application, can be found in [8] and in [9], where the ratio L/T is used instead of the controllability ratio, as well as in [5]. The implementation of the method asks for proper choices about, for instance, the amplitude of the step input and the technique adopted to fit the output response. The step response method can not be applied to those processes that are open loop unstable or that can not withstand abrupt, even if small, changes in the control effort.

The relay feedback method. The second method introduced here is the relay feedback method, introduced by Åström and Hagglund [10]. It can be classified as a frequency domain, closed loop method. When the controller is to be tuned, a relay with hysteresis is introduced in the loop in place of the controller, that is temporarily disconnected. Processes with a phase lag greater than 180° at high frequency may oscillate under relay feedback. The control variable will be a square wave, while the process output will be closer to a sinusoidal waveform. From the amplitude and frequency of the oscillation, a point on the Nyquist curve of the process with a phase shift near to 180° can be obtained. In fact, the describing function of a relay with hysteresis is:

$$N(a) = \frac{4d}{\pi a} \left(\sqrt{1 - \left(\frac{\varepsilon}{a}\right)^2} - i\frac{\varepsilon}{a} \right),$$

being d the relay amplitude, ε the relay hysteresis and a the amplitude of the input signal. The oscillation corresponds to the point where

$$G(j\omega) = -\frac{1}{Na};$$

thus the point $G(j\omega_a)$ can be determined from the amplitude a, being ω_a the frequency of the oscillation. The amplitude d and hysteresis ε have to be properly defined at implementation level. Actually d may be adjusted during the experiment to adjust the amplitude of the oscillating variables, while ε should depend on the process output noise level, that may be estimated automatically from a record of the process output. The choice of the amplitude of the relay may be more critical if the experiment has to be started from a generic non steady-state condition. The amplitude d has to be large enough that the process output overpass the set-point, otherwise the oscillation does not start. Adjustments have to be introduced also for dead time dominant processes (say for $\tau \geq$ 0.5). In this case the derivative term has no beneficial effect and a PI regulator is recommended, while the application of the design formula appropriate for time constant dominant processes may produce non-sense PID parameters (e.g., negative T_i and T_d). Also, for this class of processes, the tuning time becomes longer, compared to the average resident time of the process $(\tau + T)$. Other practical issues on this method are discussed in [11].

Gain scheduling and adaptation. Gain (or parameter) scheduling may be a very effective technique to cope with process parameters variations. To implement gain scheduling it is necessary to find measurable variables, or the setpoint, that correlate with changes in process dynamics. Then the regulator gain (other parameters) may be adjusted according to such variable values. As an example, in the control of temperature of fluids in heat exchangers, the process gain and delay frequently depend on the fluid mass flow rate, that may be fedback as an indicator of the process dynamics to adapt the regulator parameters. When the appropriate variable is available, gain scheduling may be extremely effective, and capable of following rapid changes in the process operating conditions (see, for instance, [12]). Modern PID controllers support regulator parameter scheduling. One or even all regulator parameters can be adjusted, according to an analog input signal or to the setpoint, following a piece-wise linear function assigned by the user. There are commercial products that are equipped with continuous adaptation facilities. Adaptation is obtained through diverse process modeling and identification and regulator design methods. For instance, recursive parameter estimation is used to estimate a low-order delayed discrete time model of the process. The choice of the sampling time and of a proper signal filtering is crucial in this case [13]. Other methods monitor the behaviour of the process output and adjust the controller parameters directly following a heuristic rule base. The subject of continuous adaptation is discussed extensively [5], where also the essential features of some commercial products are accounted for. according to the same authors, the use of continuous adaptation should be limited to processes, with unpredictable varying parameters, and its effective application seems to be related to loop and field devices diagnosys methods (see later on).

Feedforward control. Feedforward control is sometimes a very effective method to cope with measurable disturbances. A practical example showing excellent results is given in [12]. Modern PID have commonly an analog input port reserved for a feedforward input, that, after filtering and amplification, is added directly to the PID calculations at the regulator output.

Man-machine interface functions. The smallest 1/16DIN size controller have a quite essential operator interface, just displaying the process measured variable (PV) and its setpoint (SP) value, and the percentage of the manipulated variable (MV), possibly with no auto/manual pushbotton. On the other extremity there are the 1/2DIN, that may have a high resolution graphic panel to display trends, bargraphs, alarms, and guided menus, the latter to support selection and programming of controller functions.

Digital communications. The lower classes controllers have commonly a serial interface, based on RS232, RS422 or RS485 standards, to communicate with a supervisory computer. The computer transmits setpoint values and configuration parameters, while the controller transmits MV and PV values. Much richer are communication interfaces in high class controllers, where an RS232 port may be used for programming, and RS485 ports may be used to interface expansion or safety units, and, especially, for peer-to-peer communication with other PIDs, a DCS and PLCs. Peer-to-peer communication allows several devices to share process and control variables in real time to build up a low-cost distributed control system ("mini DCS").

Input and output capabilities. Modern controllers have direct interface with field sensors. Signal conditioning circuits support acquisition of: (1) analog inputs as Volt (0-5 or 1-5), mV (0-10-100) and mA (0-20 or 4-20) signals; (2) most popular thermocouples, like J, L, K, S, R, T, U, E types; (3) RTD and termistors, with 16 bit resolution A/D conversion. The number of analog inputs depends on the controller size: 8 is a typycal value for 1/2 DIN size, and 2 for 1/16Din size. Interface to the field comprises also isolated status inputs and outputs and analog outputs. The control output may be DC voltage, pulse, 4-20 mA, relay, triac trigger, and motor command. Two or three position outputs are pulse width modulated.

Diagnostics. Commonly controllers are equipped with self-diagnosis functions that check internal memories, A/D and D/A converters and cpu operations. Only few controllers, to our knowledge, have capabilities to monitor the operation of the entire control loop. A family of PID controllers features capabilities to detect breaks in temperature loops. Besides breaks of thermocouples (quite common), these controllers monitor other failures in the loop through additional sensors (for example, a current sensor for detecting a broken heater or a relay welded contact), and through correlation, in specific operating conditions, of MV and PV variables.

DCS

DCS are commonly used to control large plants. The first DCS was the Honeywell TDC 2000 introduced in 1975. The innovation of TDC was the integration of loop controllers with supervisory minicomputers and operator stations, based on CRT-monitors and keyboards, via a wideband digital communication bus (called data highway). DCS's have a hardware architecture that consists, at the lower level, of a number of microprocessor based modules that perform continuous variable regulation, logic control functions, and, possibly, safety functions. These modules may be geographically distributed around in the plant, and are interfaced to the process sensors and actuators through standard electrical signals (e.g., 4-20mA) or even directly. Upwards, they are linked to a local area network (LAN), through digital communication buses (data highways) and gateways. To the LAN, one ore more supervisory and processing computers, and operator and engineering stations are connected. The functions of the peripheral modules are quite similar to those of stand-alone PID controllers (also the control algorithms are the same) and of PLCs. The fast digital communications, that support the realization of a large distributed database, the computing power of supervisory computers, and the high resolution displays make a DCS a very powerful system not only for regulating the process variables, but also for monitoring and operating safely and efficiently the process. DCSs may support a true process optimization; the most popular technique to this purpose is Model Based Predictive Control (MBPC).

Model based predictive control.   "Model predictive" is said of those control approaches that embed process models. Examples of MBPC approaches are given in [14, 15, 16]. According to Froisy [17], the usual formulation of MBPC is to find a set of manipulated variables values that minimize a loss function of future predicted control errors subject to constraints on process manipulated, output, and internal variables. Control schemes of complex industrial plants have a hierarchical structure. At the lower level, that of control of ancillary equipments and actuators, PI and PID controllers are effectively used. The need, and economic convenience, of model-based control is at the higher levels, also called "money making loops" [14, 63], where optimization of the setpoints of the major process state variables, with minimization of cost functions ensuring quality and flexibility of production, can bring valuable benefits. Advantages and disadvantages of MBPC are discussed in [63]. MBPC is implemented in several commercial DCS [17, 18].

The effective implementation of MBPC require a suitable process model. Most industrial applications of MBPC use experimentally identified models, in particular discrete convolution (impulse response) models [16]. Linear parametric black-box and nonlinear first-principles models are also used. Modeling is a basic point for MBPC and, more in general, for the implementation of whatever advanced control technique. Modeling requires, and fosters, deeper knowledge of the process, that is the basis for better control, operation, diagnosis, maintanance and design (of new releases) of the process. Unfortunately, time and human resources to develop a suitable dynamic model of the process are rarely considered in the process design schedules and budgets. So, during plant commissioning, there is no way to achieve more than a simple (e.g., first order plus delay) model suitable to tune a PID. The paper by VanDoren [18] outlines an approach to modeling, adopted in a commercial product, that widely involves the user. He/she is required first to partecipate in the procedure for collecting and analysing process data, then to select the algorithms for data conditioning and for computing the appropriate model parameters, and, finally, to accept the resulting model or modify it. MATLAB commands and routines are used off line to this purpose.

PLC

PLCs [19] were originally conceived for controlling sequential logical, or discrete event, processes. To this purpose they are used in great many applications. Examples of utilisation are the control of sequential processes (for instance, in transport and material handling), problems of safety (for instance, in nuclear and in metallurgy industries), process fault diagnosis, testing of products. The most sophisticated among current commercial products are, however, upgraded with additional functions (PID algorithms) for the control of continuous processes and the command of positioning mechanisms ("axes"), so that they are a sort of DCS for small areas of automation. In consequence, they have multiprocessor modular architectures with processor, either specialized or general purpose, directed towards logic, arithmetic and floating point processing, input-output and communication.

Several languages are used for programming PLCs. Five of them are defined in the IEC standard 1131-3 [20]. Three of them, Sequential Function Charts (SFC), Ladder Diagrams (LD), and Function Block Diagrams (FBD) are graphic languages; the others, Instruction List (IL) and Structured Text (ST) are textual. SFC is intended as a specification tool, even if, being formal, it might be executed, provided that a suitable interpretation rule is declared. Formal specification tools are also the French standard language GRAFCET [21] and Petri nets [41]. The latter may support formal analysis, testing, simulation, and translation to other languages. Despite a considerable research effort is spent on the specification languages, widely used in practice are the LD, perhaps for historical reasons but also for their simplicity, ease of use and understanding.

Commercial PLCs may be directly programmed in their assembly languages, but they are generally equipped also with a PC-based CAD system that supports high-level programming.

2.2 Control devices and systems: Development trends and needs

The role of online aids to system commissioning (loop tuning and optimization) and maintenance, as well as to sensor, actuators and loop diagnostics and validation, is growing thanks to the increasing processing power within field devices and to the diffusion of a widely accepted Fieldbus standard . Better process knowledge and suitable process models seem to be essential for developments in these fields.

Autotuning and adaptation

Commissioning duration may be dramatically reduced by robust autotuning techniques. A principal difficulty in autotunig is related to the large variety of industrial processes to which PID controllers are applied. The differences may concern process dynamics properties (e.g., the controllability ratio or the open loop stability), but also generic issues related, for instance, to the possibility of operating the process in open loop temporarily, or to the user skill and will of intervening somehow in the tuning process, or to the admissible duration of the tuning procedure (compared with the process dynamics), or to the apriori available information. There exists no autotuning method suitable for all situations. Each manufacturer, therefore, selects and implements in his products the method most suitable for his target customer. A method able to handle satisfactorily a very wide process family, i.e. robust with respect to process nature variations, is not available yet.

Currently continuous adaptation techniques are rarely applied. Åström and Hagglund [5] point out that poor controller tuning and troubles in field devices (for instance, strong increase of stiction in a valve actuator due to wear) may both manifest in process output oscillations, but they likely require quite different corrective actions: loop (de)tuning the first case, maintenance the second one. Thus, to apply loop adaptation effectively, it seems desirable to supply the controller with on line methods of sophisticated diagnostics of field device behaviour. Research in this area is rapidly growing (see also later on).

Another difficulty in continuous adaptations is related to time varying process delays, since, when the time delay appears as a parameter to be estimated, the model to be identified becomes nonlinear in the parameters, even for a linear process. Several methods have been proposed to afford the problem [22]; nevertheless a truly satisfactory (that is able to tackle a large variety of situations) approach has not been arranged, yet.

The Fieldbus

Currently, DCSs, PLCs and PID controllers are usually connected with sensors and actuators through analog 4-20 mA transmission signals. Even the modern smart field devices, with internal digital processing, are equipped with a D/A converter for analog interfacing. However, the need for a digital replacement of the 4-20 mA signal is strongly felt by manufacturers and users, and several manufacturers have developed, since mid 1980s, proprietary fieldbuses. The fieldbus was originally intended to replace the analog, point-to-point communication, with an open, multidrop digital communication system (on the subject of fieldbus see the Fieldbus Series in Control Engineering [23]). Even if many different types of wires can be used, a simple twisted pair of wires suffices to connect many (say 32) devices. Reducing cabling requirements is the first short-term benefit of choosing fieldbus based communications. This benefit is less obvious and ample than expected if risks of loss of sensors and control is considered; nevertheless, even considering reasonable requirements of redundancy, there are potential advantages in terms of lower initial costs and of lower maintenance costs.

The most important, and perhaps revolutionary, expected benefits are, however, related to the two-way communication between the controllers and the field devices, that is feasible with fieldbus. Potentially this may support online diagnostics, predictive maintenance, easier calibration, easy access to detailed device information, and improved control system performance.

Several international organizations and suppliers' working groups have been working for many years to define standards for the fieldbus. Users push for the definition of a single, unified, interoperable fieldbus. Interoperability is the essential long-term benefit of fieldbus. It is intended as the possibility not only to interconnect devices from different suppliers, but also to have them working together in a transparent way for the user, with the possibility of replacing a device from a supplier with one of a different supplier without restriction on the function the device provides. Interoperability requires standardization of devices' interfaces, common data services and functions. The standard currently under definition by the Fieldbus Foundation (created in '94, by merging the two major organizations in North America, the InterOperable Systems Project Foundation - ISPF - and the WorldFip North America) consists of four levels, defined in agreement with the OSI (Open System Interconnect) reference model, and of the network/systems management services. The layers are, from the lowest one:

- The Physical Layer (PL), encoding and decoding data on the transmission medium;

- The Data Link Layer (DLL), providing services for transparent and reliable data transfer among users;

- The Application Layer (AL) providing the services required to support process control and factory automation applications to the User Layer;

- The User Layer (UL) running user defined applications.

When an agreement on the definition of the standard for the UL is reached, the implementation of true open and interoperable control systems will be possible. The function of the UL is to allow suppliers to build measurement and control products with predefined algorithms,

easily assembled by the users to define their specific control application. The UL has to completely define the data acquisition and control
functions of each field device. It defines all control blocks necessary to
implement a generic control system and the structure of the database associated to each block. In this way the "external" behaviour of a device
on a fieldbus will be known to the others, even if the internal implementation of its functions will be specific of each supplier. Thanks also
to fast, two-way communications among devices, a downward shift will
be obtained in control tasks within the automation hierarchy, with simple control tasks really migrating into the field, and with the creation
of a widely distributed database. Each device should include specific
information for programmed maintenance and online diagnostics (refer
also to the PRIAM Project of the ESPRIT Program [24]). Obviously,
the PID still remains a fundamental functional block, though, according to Kompass [25], the fieldbus-based distributed architecture will, in
the long term, seemingly generate revolutionary innovations in control
techniques.

Field device validation and diagnosis

Advances in microelectronics and the development of (proprietary) digital communications has fostered the recent development of digital sensors with multiple sensing capabilities (for instance, they jointly measure
fluid flow, temperature and pressure). Digital devices, together with the
Fieldbus, are also fostering the development of techniques for field devices and loop validation. It is observed in [26] and [27] that in-built
processing power can exploit device specific knowledge and online information, to allow the generation, in addition to a validated measurement
value, of a validated uncertainty, of the measurement value status, and
of a device status. This information can be made available, through the
Fieldbus, to the process controllers and supervisor so that the proper
actions are taken. The research on device validation, fault detection and
identification, as well as accomodation, is still in its infancy [27].

Model based predictive control

Froisy [17] underlines that achieving process models, the foundations
of model based predictive control, will continue to be the most difficult and time consuming task for practitioners. Thus, widespread applications will require cost-effective dynamic modeling platforms with

suitable modeling tools and computing power. Effective modeling will support the implementation of general nonlinear controllers. In this view, attention is paied also to neural networks. MBPC likely will benefit from improvements in DCS, digital communications (Fieldbus), and sensor technology as well.

3 System Engineering: Advance Practice and Trends.

Let us consider a simple example taken from a typical control engineering task: the design of the control of the feed-water system of a steam plant.

The input of the control system engineers starts with some process flow diagrams like that of Fig. 1 (which describes the overall process organization) and with a process control philosophy stemming from the operating requirements. Data available to control engineer concern piping, pump's characteristics, valve flow condition, heat exchanger and storage tanks sizing, nominal process conditions and expected range of operation. In addition, a set of operating constraints, safety conditions and procedures are usually specified.

All the above data and others are (or should be) stored in a project data-base, where information from all engineering disciplines (at least) is shared-data being deposited by the discipline responsible for generating it, and withdrawn by the discipline which needs it. In our example, the data that the control engineer is asked to supply are related to the following tasks:

1. specification of control system objectives, in detail;

2. specification of instrumentation (e.g. sizing of control valves);

3. preparation of control system schematics, both for the modulating control part and the logic control part;

4. selection of control system parameters;

5. specification of acceptance tests for the final system.

Often it occours that, in doing that, the system designer has in mind a specific control system vendor for the instrumentation (especially for DCS and PLC) but it is preferred that the design could be vendor-independent, to enhance project flexibility.

There is a constant push to reduce engineering and design costs, design cycle time, design errors, duplication of work and to automate engineering work processes. Integration of project data and vendor-independent specification are among the principal means to move in this direction.

Unfortunately, in the field of control there are still many aspects for which the description in terms of data is not given a standard format. In particular, the above-listed task (1) is not covered by any systematic methodology (to formalize control system objectives), particularly for process plant applications, so that control objectives are very often specified in terms of (possibly ambiguous) statements, such as: "the feedwater system has to allow feedwater control within the range of 10%-100% of the nominal load"; or "pump # 3 will automatically start in case of trip of either pump #1 or # 2". And even in those cases where more precise performance specifications are defined, in terms of times, event sequences or control errors, there is not any engineering procedure established to ensure that the design will meet the specification. In this respect, interesting tentative proposals to supply the design with a systematic approach to objectives specification have been put forth in the area of aerospace applications by NASA [28] and by ESA [29].

The basic concept of those proposals is to fix a *Reference Model* in a given application field, with the aim of using it to standardize the whole Project life-cycle and to guarantee traceability of the project development. Strict association is pursued between a certain control task and a "typical" control module in the *reference model* architecture.

This sems a good suggestion also for industrial applications, but still quite far from engineering practice.

Specification of instrumentation has already achieved a high level of standardization (refer, for instance, to ISA, DIN or IEC[1] standards), so that instrumentation data-sheets are usually (or could be) embedded in the project data-base as shared data. Take, as an example, the work process involved with sizing a control valve [30]: process data are transferred from a process engineer into the common data-base from which it is withdrawn by the control system engineer, who runs the relevant sizing program, sizes and specifies the valve, and return size and dimensional data to the common data-base; this data are available for subsequent operations (e.g. done by the piping engineer). Because data

---

[1]ISA: *Instrument Society of America*; DIN *Deutsches Institut für Normung*; IEC: *International Electrotechnical Commission*.

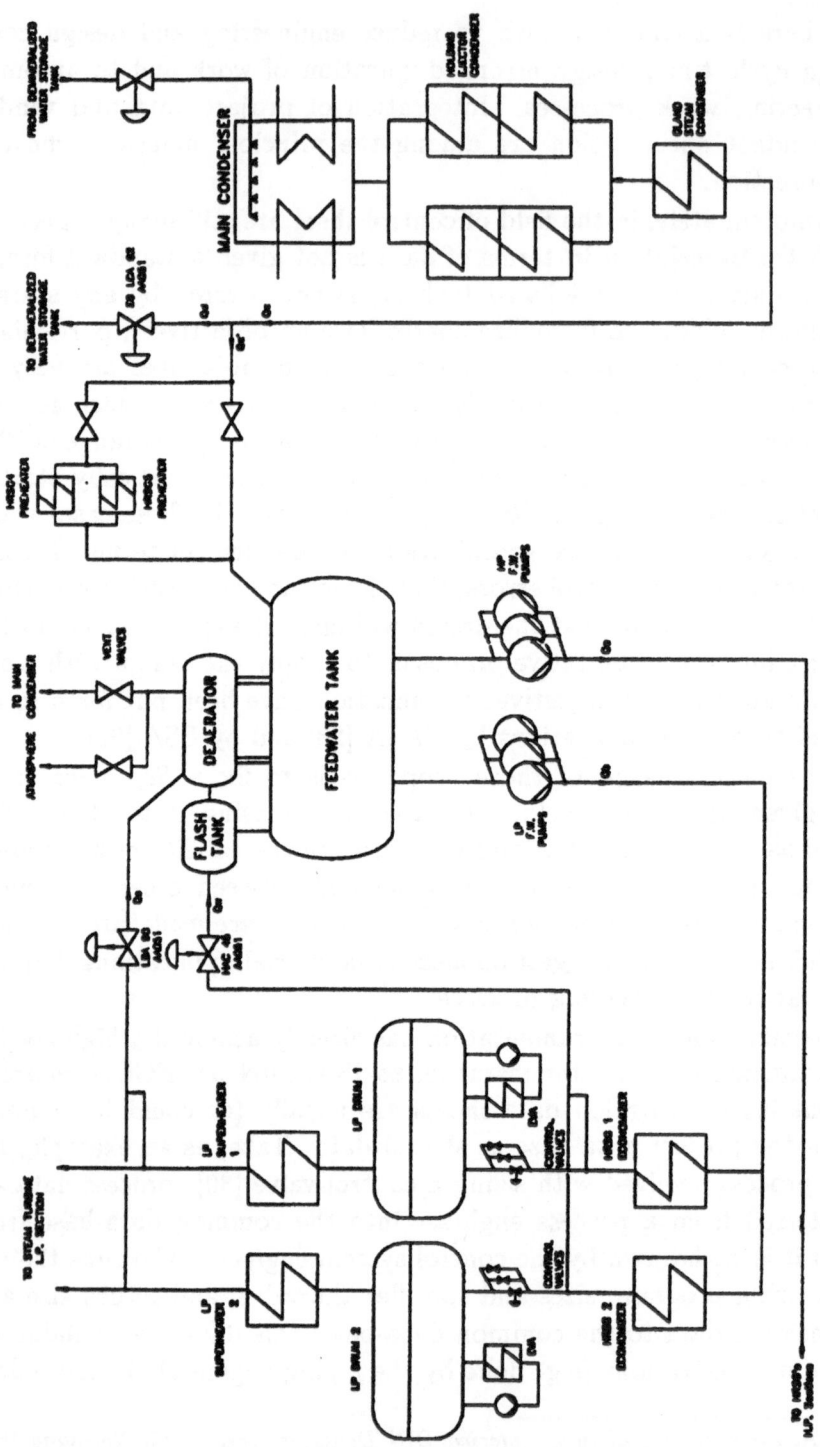

Figure 1: Process flow diagram of the feed-water supply system.

are generally used in different forms by engineers of different disciplines, it is often convenient, to support data storing and retrieving, to use an expert system that acts as an interface between the users and the data-base.

The introduction of this methodology has impressively improved the quality and productivity of the instrumentation engineer.

Control schematics are generally prepared using an international standard (typically ISA, DIN or IEC); large companies sometime use a proprietary symbol dictionary, which is generally a convenient customization of some international standard allowing better specification.

The development of control schemes are mostly based on experience, i.e. the control engineer builds control "pseudo-programs" in the form of block-diagram making reference to typical subprograms already applied to similar problems. However, because of the very little modeling power of the international standards (as compared with the algorithmic potential of digital control systems), there is a real push to extend the set of standard control elements to incorporate more complex programs as macros.

Unfortunately, as is asserted in [31],"rather than simplifying the life of the control engineer, this has tended to increase the problems of implementing complex control strategies and achieving benefits from then".

The problem here is that, while Distributed Control Systems (DCS) are widely used powerful platforms for plant-wide control, there is no international recognized standard for the algorithm set of a DCS. So, increasing the software complexity of DCS reduce the understanding of the software real behaviour by the control engineer and his possibility to exploit the DCS potential. On the other side, the international standards for Process & Instrumentation Diagrams (P&ID) do not fully define the algorithm implementation; in other terms, they are ambiguous because they describe rather control concepts than control algorithms, so that it is not always possible to automatically build control (executable) programs from diagrams.

The above remarks apply both to modulating control and to logic control, in the latter case especially when sequential functions are described.With reference to the process of Fig. 1, control schemes are specified, for example, as described in Fig. 2 and 3 using a symbol set taken from DIN standards.

In engineering practice, to overcome the above mentioned ambiguity of the control symbols set, both engineering companies and DCS vendors

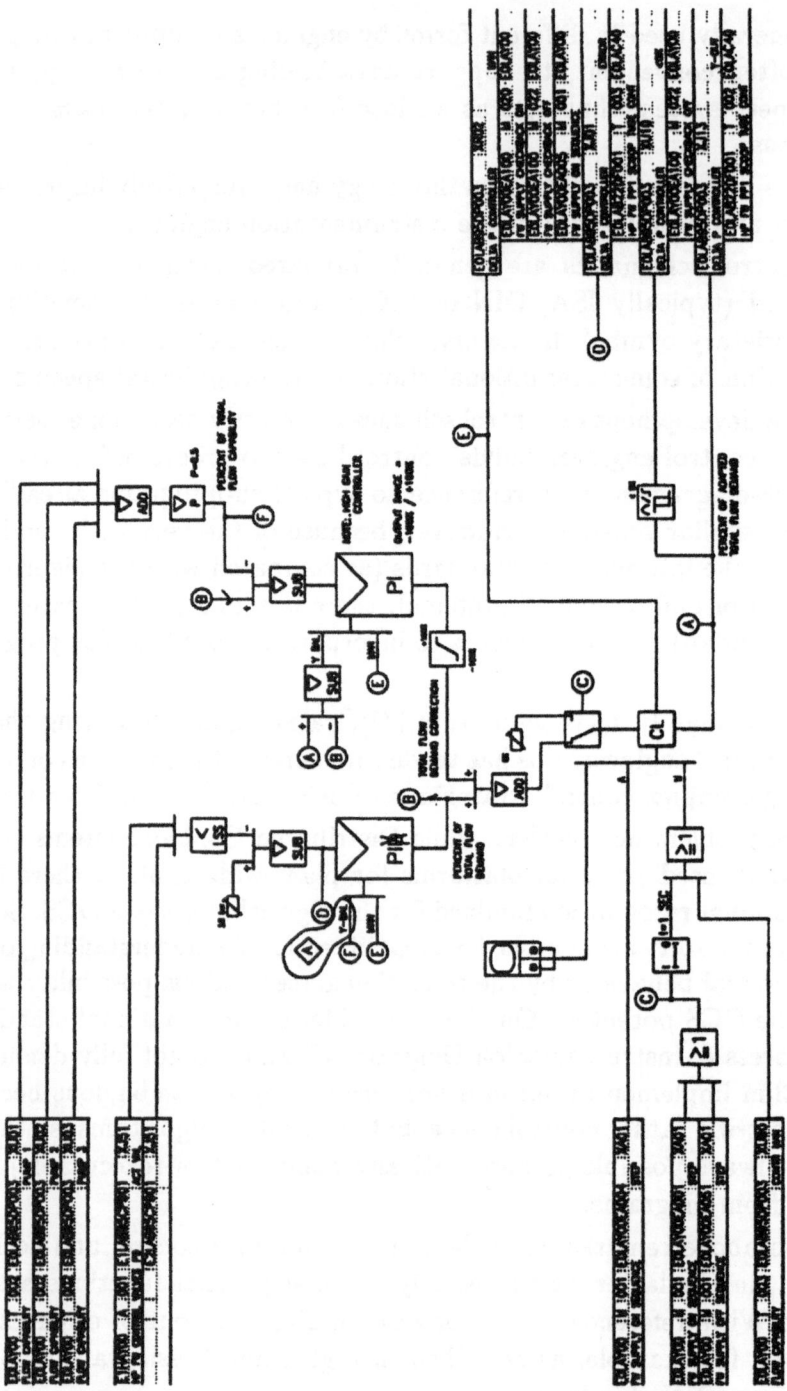

Figure 2: Differential pressure control scheme relative to the high pressure feed-water control valves (master controller).

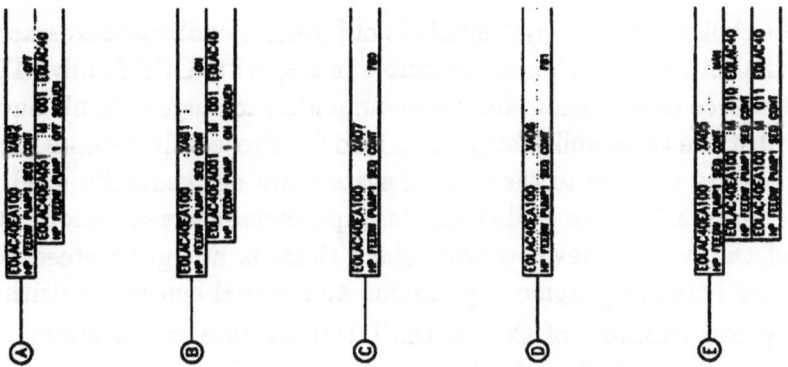

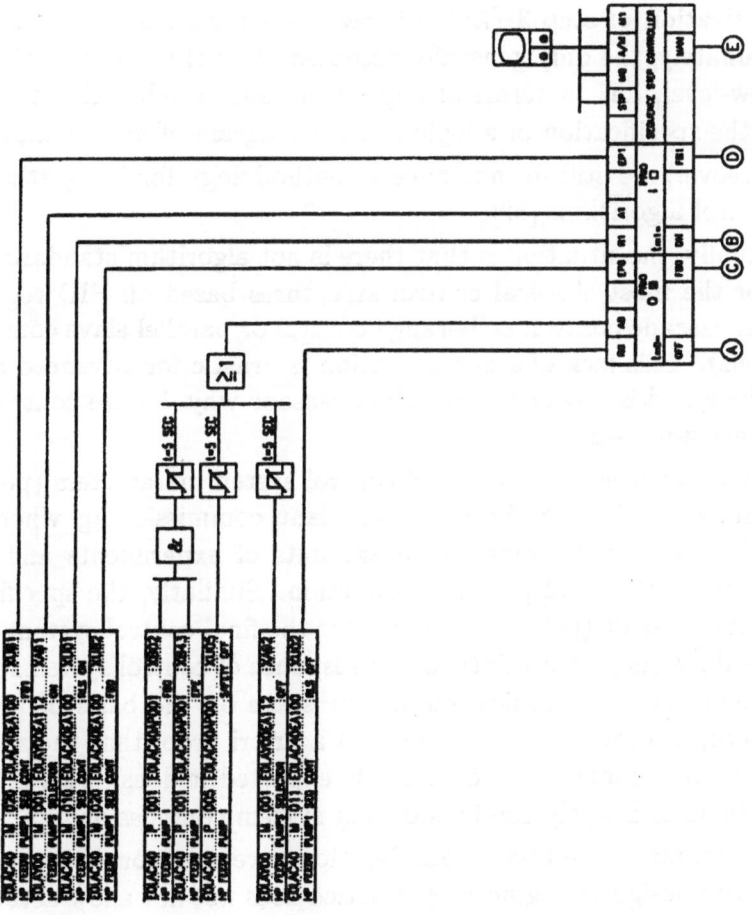

Figure 3: Logic diagram of the Sequence Start relative to a feed-water pump (any step in the sequence is specified by a dedicated separate logic diagram).

often specialize the "generic" symbols of international standards according to the actual algorithm set available in a specific DCS family. This is a great source of confusion, makes documents exchange difficult and is a severe obstacle to establish standards also for Project data-bases. In any case, the data relative to the control system are not generally used (and usable) for quantitative evaluation (e.g. performance evaluation) at the stage of the project development, since there is not guaranteed correspondence between graphic expressions and actual control algorithms.

A typical example of that is the interpretation of the simple logic control diagram of Fig. 4, showing the second step of the sequence activated by he diagram of Fig. 3. The description is "conventional": for instance it is intended that if the ENabling Conditions (ENC) of step 3 hold, then step 2 is jumped; moreover it is assumed that, in some way, activation of step 3 (ACT) is reset when step 3 is executed, etc. Unfurtunately, the only generally accepted algorithmic description is at very low-level, i.e. in terms of logic ports and flip-flop elements, that makes the specification of a logic control program of huge complexity.

Moreover, we still do not have a methodology for integrating logic and control algorithms [32].

Actually, the situation is that there is not algorithm standardization even for the most classical control structures based on PID regulators (such as cascade-control, split-range control or parallel slave control applications). This lack of standardization is drastic for advanced control algorithms and is one of the principal reasons why diverse control algorithms are not used.

As a consequence, selection of control system parameters (point (4) listed above) is left to the phase of plant commissioning where controllers' tuning may require huge amounts of experiments and imply quite important loss of process production. Similarly, the specification of acceptance tests (points (5) above) for the final control system is subject to substantial uncertainty as the response of control systems is only "presumed" at the specification/design stage on the basis of previous experience; in other terms, there is not a priori proof that the specified control system could really do what is expected it does. This is source of fastidious and costly conflicts during system commissioning.

To summarize the above considerations, we may conclude that control system design in engineering practice does not not allow actual performance evaluation but only the "approximate" prediction of the "presumed" behaviour on the basis of available experience on similar plants.

In fact, computations essentially regards only control instrumentation sizing but not global system response. Consequently, control design usually "follows" the other design choices involved in a given project and there is *no integration* of control with the other disciplines during design: for new process concepts this may imply dramatic mismatch between process design and achievable process operating performance, especially when process dynamic features, crucial for actual control performance, are ignored while sizing the process (refer to [12] for a very expressive example of possible process-control mismatch that may occur without integrated design).

The basic need is to conceive the project of an industrial plant as a fully integrated activity and the engineering work (including control) as a "concurrent engineering" process. So we have to ask what are the lines along which such requirement can be accomplished.

The goals are to remove ambiguity in system representation: this means introducing better standards and adequate methodologies and tools to support reliable prediction of system performance. The crucial role, in the author's opinion, will be plaied by advanced CAE systems, capable to deal also with Control Engineering, but with a careful consideration of integration. The idea is to look at a CACE tool organically linked to the Project data-base, with data exchange possibly aided by a knowledge based system (see Fig. 5 and [33]).

Proposal of integration of multiple tools and of multiple data-bases to implement full integration of CAE activities are the subject of various research (see e.g. [34]). Refering to CACE concepts, it is generally meant [35] that "CACE is a specialized form of modeling and simulation with emphasis on design and implementation of feedback control systems". To complete this definition is should be also added "and with system representation respecting engineering formats and standards". As such, the core of any CACE environment is modeling and simulation, where the physical part of the system is decribed according to current process structuring principles (refer e.g. to Fig. 1) and the control & instrumentation part is described according to (possibly improved) control engineering standards (refer to Figs. 2, 3 and 4).

The use of engineering simulators to significantly improve control design and implementation in large engineering projects has already been proved in many real applications (for instance reference [36] describes a very significant case in the area of power plants). Yet, there is still a substantial gap to be filled in order to make such an approach really ap-

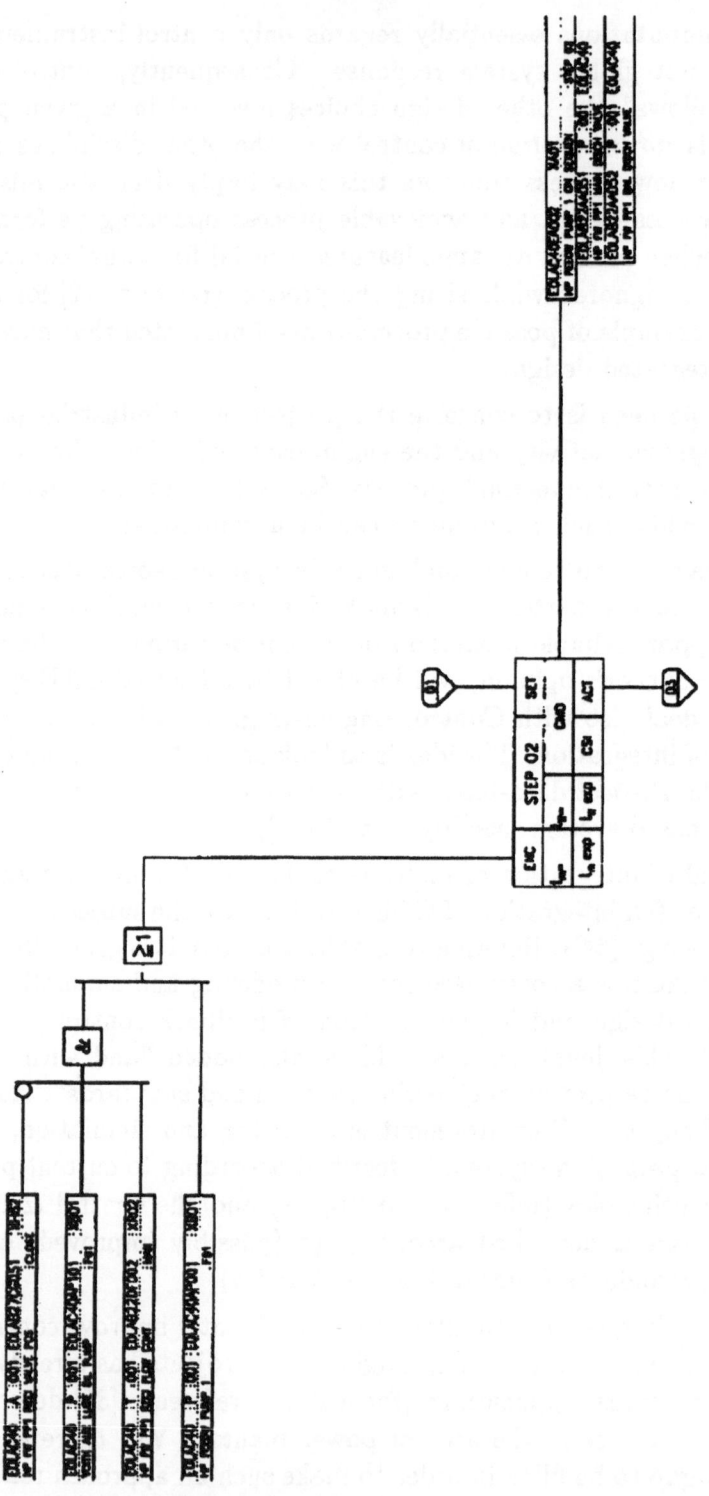

Figure 4: A step of the pump start-up sequence.

plicable in engineering practice, where a project must have a "normal" life cycle and a definite completion time.

Once again, standardization is a prerequisite. In this direction, principles of engineering modeling and simulation have recently been introduced by refering to *object-oriented* model structuring [37, 38, 39]. The object-oriented approach is considered as the best way to match the natural engineerig description of physical systems by components. Much progress has been done on crucial mathematical aspects (takling complexity with large DAE systems) and on software aspects (prototype languages for modeling [40] and use of object-oriented data-bases [39]); also the methods to deal with hybrid systems [32, 41] have been improved and clarified. However, the crucial point of standardization (of models' and submodels' interfacing, of simulation techniques for complex systems, of hybrid system representation) is still far from being satisfactorily solved.

Since engineering simulation based on "typical templates" is the only realistic way to pass from system design by experience to system design by direct evaluation, the role of simulation/simulator toolboxes conceived for different engineering domains is becoming crucial for any step forward in control engineering. The task involved is certainly bejond the possibility of a simple reasearch group; so the *European Science Foundation* (ESF) is funding a Program (called *COSY* Program) on Control of Complex Systems, to put together the efforts of different reasearch groups in Europe, with the aim of developing pre-competitive actions in the fieds of CACE software.

We believe that modeling and simulation environments specifically conceived for engineering and suitably integrated in the project database is the today challange to port control design methods outside academia. To do this, we need guidelines to develop simple and intuitive modeling methods, both for physical systems and for control systems, meeting the engineering requirements of specification and standardization.

Problems strongly related to modeling are the design of plant-wide control strategy and the interaction between process design and process control. In current engineering practice, control strategies are generally conceived on the basis of experience and of heuristic (static) considerations. Even well known methods of aiding inputs-outputs pairing (e.g. the Bristol's Relative Gain Array (RGA) [42]) or to asses process controllability in dependence on the control structure (refer to [43] and to the paper referenced there) are not applicable during the normal life-cycle of an engineering project because their application needs fast-development modeling plantforms. Static modeling (refer again to [43] may be useful, but sometime misleading, in those cases where process dynamics determines process operability (refer, for example, to [12] and to [44]).

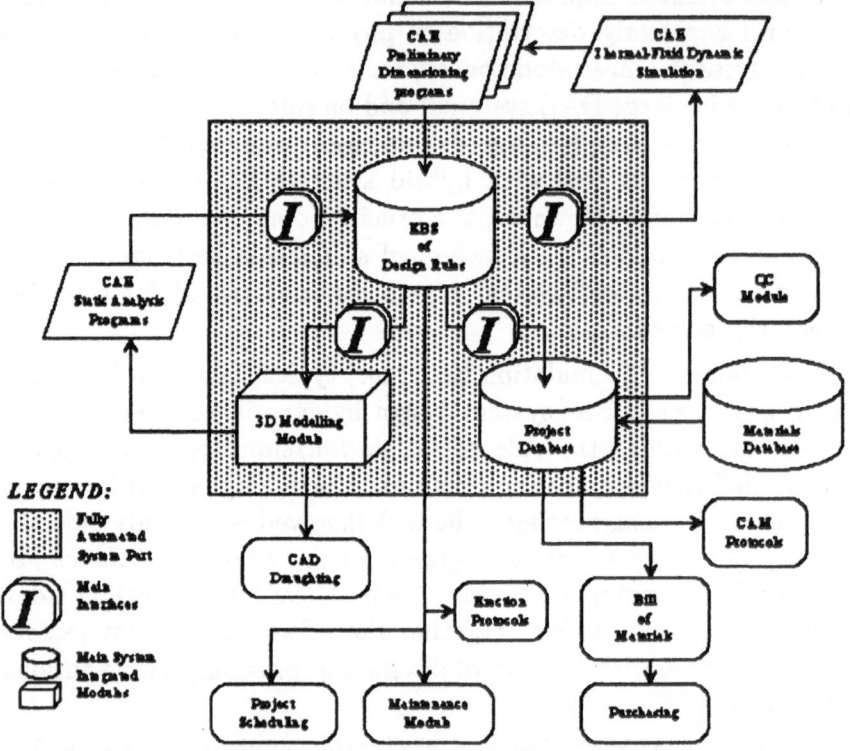

Figure 5: Integration of CACE tool in a CAE environment.

Moreover, available multivariable control methods, based on a pure mathematical representation of the probelm at hand, are not conceived to account for very important practical issues, such as tolerance to measurement failure or drift, understandability by the plant operators, tunability by simple field experiments, hardware/software architectural considerations, maintenance, accuracy and reliability of disturbance measurements, etc.. A real-size case study is illustrated in [45], that shows how the multivariable structure of the process can be dealt with by a control system consisting of classical PID controllers suitably arranged and tuned according to the process dynamics and constraints.

In conclusion, we can say that the possibility of applying dynamic performance considerations to design the control strategy is still quite limited in practice and, in any case, integration of control methods within the engineering work process is a crucial bottleneck not yet removed neither in theory nor in practice. It is expected that powerful CACE plateforms could overcome this serious obstacle.

4   Role of Control Science in Control Engineering Progress.

In the last ten years, control engineering has significantly increased its importance in the engineering contractor environment, especially because (see [43]) "more efficient and environmentally-friendly production leads to processes with more energy and material recycling and hence more complex dynamics". However, we may claim that the impressive production of control scientists has not significantly affected the technology used and the way to use it in engineering projects. This is particularly true for the sophisticated and wide area of multivariable systems control, including control strategy design methods (recalled in the preceding paragraph) and synthesis of linear and nonlinear controllers. Even in the peculiar case of robotics, where systems are multivariable and nonlinear and where the theory of feedback linearization [46] seems to fit quite well the problem, industrial robot controllers do not generally apply such concepts.

We could say that only a few percents of the control science results obtained in the last thirty years have been applied to new products or to new methodologies for executing projects, particularly if we make reference to industrial products and processes. The general reasons why this has occurred are, at least in part, discussed in the previous section and could be summarized in the statement that, unlike in the fileld of software, research in the field of automatic control has not developed a specific discipline of "control engineering".

4.1   New products

Looking at the development of new products in the recent past, it can be recognized that ease of tunability and simplicity of use have been the key properties of success innovations. In this line, there is a strong expectation concerning *fuzzy control* techniques [47] and neural networks [48], considered as ways to avoid accurate analysis and complex modeling.

The potential of such techniques for the simplification of the engineering work has not yet been analysed with sufficient criticism even though new products already appeared on the market implementing fuzzy logic and neural networks.

The most interesting feature for engineering is that fuzzy logic and neural networks are intrinsically oriented to standardization; scopes and limits, however, are still far from being clarified and qualified for applications.

Very important research lines for their practical impact are those concerning *industrial diagnostics* [49] and *loop validation* [27], especially because the emerging technology of the Fieldbus makes available relevant additional information about sensors' and actuators' state and relevant intelligence aboard instrumentation. Since plant availability is the first issue in practice, methods to improve detection, diagnosis and advance maintenance will be of great economic benefit for plant operation. Here, control research should address the problem of how to develop prototypes capable of complying with control system architectures, either as part of intelligent instruments or as part of the plant DCS.

Industrial control systems are presently not equipped with significant capability of takling multivariable processes; well proved methods such as *Model Predictive Heuristic Control* [14], *Internal Model Control* [50] and *Dynamic Matrix Control* [15] are not given generalised support in industrial systems, though they are used, as background tools, for case studies, to devise improved control strategies. What is still needed is a specific effort to engineer most effective methods, in order to bring them into the control "toolbox" available in the process computer. Since the control engineer can not develop case studies during plant commissioning, algorithm set-up must be reduced to "tuning" a limited number of easily understandable knobs. This requires further, non trivial, dedicated research.

4.2   Methods and tools for (Computer Aided) Control Engineering

There is a ripe demand from industry of plant-wide *simulator toolboxes*, with such features as to allow simulation to be included in the "normal" life-cycle of a project development. Research has to study not only new abstract concepts to improve model structuring, but also basic requirements and constraints coming from project development prectice, where data integration, concurrent engineering and project timing are prerequisite for any working tool. We believe that neutral formats [51] have to be combined with dedicated libraries and data structures fitting specific domains of application, as in the case described in [52].

Moreover, to follow the different phases of project development (from feasibility study to commissioning), simulator toolboxes should be flexible in use and suitable for incremental building, with regard both to increasing detail and to subsystems' aggregations. It has to be underscored that commercial mathematical packages (like *Simulink/Matlab^{TM}*) or general purpose simulation languages (like *ACSL^{TM}*) are not applicable within engineering project contracts, because they require a system description in terms of equations and not of (design) data. This is a real bottleneck between control engineering and simulation, that is between control engineering and quantitative evaluation of system performance.

Research follows three main lines:

- improving system representation in terms of components and data, typically according to the object-oriented approach (see e.g. [38, 39, 40]);

- defining more powerful standards for control systems description, allowing integarted representation on control algorithms and of system architecture, in order to obtain realistic predictions of system performance in real-time (see e.g. [41] and [53]);

- developing environments/libraries specialised in particular application domains, such as *Mechatronics* [54], *Multibody Systems* [55], *Power Plants* [52] or *Chemical Processes* [56].

Much work is done and is still required to integrate modeling and simulation platforms with control analysis and design toolboxes. Various prototypes have been developed with this purpose; among these ones the *Process Controllability Toolbox* [43], *ANDECS* [34], *CAMeL* [57] and *Dynamics-Control* [55] are significant examples.

The use of *Control Development Methodologies* and of *Reference Models* [28, 29] guiding control system structuring is very important to establish systematic procedures in the specification/design process; real-size examples of application are reported in [53] and [58] with reference to robotics. However, the approach seems really to be at a much earlier stage with respect to similar proposals in the field of *Software Engineering*. Extension of Reference Models to different application fields (e.g. process contro) and integration with more traditional CACE frameworks is expected for the next future.

4.3   On-line additional functions

Thanks to the capability of process computer (network) technology, a number of real-time functions have now become feasible, that can be seen as an extension of traditional control functions.

In a paper of G.K. Lausterer [59] of Siemens AG (Power Generation KWU), the key issue of the economic benefits of advanced control is considered with reference to power plants. After having considered what are the principal areas of possibble economic return in power plants (e.g. efficiency, lifetime, maintenance etc.), some measures, based on *Instrumentation & Control*, are illustrated that have been successfully applied (with economic benefit) in power plants. Beside improved feedback control and new control strategies, real-time and predictive models are extensively exploited to perform the following on-line functions:

- *predictive trending* [60], that is continuous prediction of next-future evolution of relevant process quantities, to warn the operator about possible alarm conditions;

- *load margin calculator*, that acts as an *operator guide* during start up, load variations and other manoeuvres, showing the actual plant operating margins at any time based on the current state;

- *optimization computer* [61], that determines the best arrangement of the process operation to achieve, at any time, best global efficiency;

- *knowledge-based* operator aids [62], for monitoring and on-line diagnosis of correct/incorrect plant operation.

The actual return of that kind of additional on-line functions has already been proved either in continuous operation (for example, functions similar to the load margin calculator are already operating in commercial plants) or with extended experimental testing. They are based on different combined technologies, such as *"what if"* simulators, obsevers, expert systems, predictive control and quasi-static optimization. Treir role in the comprehensive design of a real control system is expected to increase significantly, because computing costs are quite low and network-based distributed systems are drastically cutting the cost of wiring for measurements.

However, the available methodologies are mostly heuristic; control engineering research is required to give them a more rigorous assessment and a better correspondence to the operator's way of thinking.

5 Conclusions

We have discussed topics useful to build a bridge between control science and control engineering, starting from the advance practice in control engineering and considering its fundamental requirements.

It has been recognized that engineering project data integration and improved standards are of key importance for the improvement of the *professional work process* and for the mere possibility of introducing better methods in control engineering.

The development both of new products and of new design procedures are only slightly influenced today by control science, in the sense that the results made available by control science do not generally fit most of the engineering constraints and requirements. There are fortunate exceptions to this general condition, as illustrated in Section 2: they are cases where specific (non-trivial) work has been done to comply with the practical use of new concepts.

In the field of control algorithms, only PID control is close to a satisfactory standardization, including the simplest autotuning algorithms. More advanced control algorithms and even complex PID-based control structures are only available as proprietary software modules.

CACE frameworks seem to be candidate platforms to implement the integration between standards and documentation required in engineering project development, on one side, and analysis and design methods as conceived by control science, on the other side.

Finally, process computer and disributed instrumentation & control systems allow the implementation of new effective on-line functions, supplying important aids to the operator to improve plant efficiency, minimize plant outages, reduce maintenance costs, etc.. Much attention is paid by engineering contractors to the wider area of control & instrumentation, considering both automatic control in strict sense and advanced operator aids, such as monitoring, simulators, expert systems, etc..

Looking at the current state of the art, we may conclude that control science, very tightly linked to developing new sophisticated methods for feedback control analysis and synthesis, is not sufficiently motivated and oriented by control engineering expectations to determine significant impact on control engineering progress.

6   References

[1] Boyette P., Bhullar R. S. Challenges of tomorrow for the control systems professionals. ISA Transactions 1994, vol. 33, pp. 197-205.

[2] Taylor J. Temperature controller market and product trends. Control Engineering, Nov. 1993.

[3] Morris H. Multiloop DIN-sized process controllers keep getting smaller. Control Engineering, Feb. 1993.

[4] Babb M. New multifunction controller provides advanced strategies. Control Engineering, Sept. 1995, pp. 57-58.

[5] Åström K., Hagglund T. PID Controllers: theory, design and tuning. Second ed., ISA, 1995.

[6] Vandoren V. Inside self-tuning PID controllers. Control Engineering, August 1993.

[7] Ziegler J.G., Nichols N. B. Optimum settings for automatic controllers. ASME Transactions, Vol. 64, 1942, pp. 759-768.

[8] Hang C. C., Åström K., Ho W. K. Refinements of the Ziegler-Nichols tuning formula. Proc. IEE, Pt. D, 138, no. 2, 111-118, 1991.

[9] Åström K., et alii. Towards Intelligent PID Control, Automatica, No.1, 1992.

[10] Åström K., Hagglund T. Automatic tuning of simple regulators with specifications on phase and amplitude margins. Automatica, Vol. 20, No.5, 1984.

[11] Hang, C. C.,Åström K. Practical aspects of PID auto-tuners based on relay feedback, Prepr. IFAC Symp. on Adaptive Control of Chemical Processes, ADCHEM '88, Lyngby, Denmark, 1988.

[12] Maffezzoni C., Magnani G. A., Quatela S. Process and control design of high temperature solar receivers: an integrated approach. IEEE Transactions on Automatic Control, vol. AC-30 pp. 194-209, 1984.

[13] Ferretti G., Maffezzoni C., Scattolini R. On the identifiability of time delay with least squares methods. Automatica 1996, vol. 32, no. 3, pp. 449-453.

[14] Richalet J. A., Rault A., Testud J. L., Papon J. Model predictive heuristic control: application to an industrial process. Automatica 1978, vol. 14, pp. 413-428.

[15] Cutler C. R., Ramaker B. L. Dynamic matrix control: a computer control algorithm. Proc. Joint Autom. Control Conf., San Francisco, 1980.

[16] Seborg D. E., T.F. Edgar, D.A. Mellichamp. Process dynamics and control. Wiley Series in Chem. Eng., 1989

[17] Froisy J.B. Model based predictive control: paste, present and future, ISA Transactions 33, pp. 235-243, 1994.

[18] Vandoren V. Advances in model-based control technology. Control Engineering, Sept. 1994.

[19] Olsson G. Programmable controllers. In: Levine W S (ed), The Control Handbook, CRC Press - IEEE Press, 1996, pp. 345-361.

[20] International Electrotechnical Commission. International Standard IEC 1131-3, 1st Edition, 1993.

[21] David R. Grafcet: a powerful tool for specification of logic controllers. IEEE Trans. on Control Systems Technology 1995, vol.3, no. 3, pp. 253-268.

[22] Ferretti G., Maffezzoni C., Scattolini R. The recursive estimation of time delay in sampled-data control systems. In: Leondes C (ed.), Control and Dynamic Systems, vol. 73, Academic Press, 1995.

[23] Fieldbus Series, Control Engineering, January, March, May, July, September, December, 1994.

[24] Capetta L., Galara D., Graefer J., Lovishek G. P., Mondeil L., Sanguineti M. From current acrtuators and transmitters toward intelligent actuation and measurement: PRIAM approach. Proc. of International BIAS Conf., Milan (Italy), 1993.

[25] Kompass E. A 40-Year perspective on control engineering. Control Engineering, Sept. 1994.

[26] Henry M.P., Clarke D. W. The self-validating sensor: rationale, definitions and examples. Control Engineering Practice. 1(4), 585-610, 1993.

[27] Clarke D. W. Sensor, actuator and loop validation. IEEE Control System Magazine, Vol. 15, 4, Aug. 1995.

[28] Albus J. S., Quintero R. Toward a reference model architecture for real-time intelligent control systems (ARTICS). In: Robotics and Manufacturing, vol. 3, ASME Press, 1990.

[29] Putz P., Elfving A. ESA's control development methodology for space AR systems. In: Jamshidi M, et al. (eds), Robotics and Manufacturing: Recent Trends in Research, Education and Appl., vol. 4, ASME Press, 1992.

[30] Faccini G, Tira P. Roles and tools of engineering contractors. Automazione e Strumentazione Nov. 1995, pp. 107-113 (in italian).

[31] McCulloch G. Reduced algorithm set control. Control Engineering Practice 1996, vol. 4, no. 2, pp.195-205.

[32] Kohn W., James J., Nerode A., Harbison K., Agrawala A. A hybrid system approach to computer-aided control engineering. IEEE Control Systems 1995, vol. 15, no. 2, pp. 14-25.

[33] James J. R., Taylor J. H., Frederick D. K. An expert system architecture for coping with complexity in computer-aided control engineering. Proc. 3rd IFAC/IFIP Symp. on Computer-Aided Design in Control and Engineering Systems, Beijing (PRC), 1988.

[34] Grübel G. The ANDECS CACE framework. IEEE Control Systems 1995, vol. 15, no. 2, pp. 8-13.

[35] James J., Cellier F., Pang G., Gray J., Mattsson S. E. The state of computer-aided control system design (CACSD). IEEE Control Systems 1995, vol. 15, no. 2, pp. 6-7.

[36] Groppelli P., Maini M., Pedrini G., Radice A. On plant testing of control systems by a real-time simulator. 2nd Annual

ISA/EPRI Joint Control and Instrumentation Conference, Kansas Cyty (USA), 1-3 June, 1992.

[37] Cellier F. E., Elmquist H., Otter M. Modeling from physical principles. In: Levine W S (ed), The Control Handbook, CRC Press - IEEE Press, 1996, pp. 99-108.

[38] Mattsson S. E., Andersson M., Åström K. J. Object-oriented modeling and simulation. In: Linkens D A (ed.), CAD for Control Systems, Marcel Dekker Inc., 1993, pp. 31-69.

[39] Maffezzoni C, Girelli R, Lluka P. Object-Oriented database support for modular modelling and simulation. Proc. European Simulation Multiconference ESM '94, Barcellona, June 1-3, 1994.

[40] Andersson M. Object-oriented modeling and simulation of hybrid systems. PhD Thesis, Dept. Autom. Control, Lund Inst. of Technology, Lund, Sweden, Dec. 1994.

[41] Ferrarini L., Maffezzoni C.A conceptual framework for the design of logic control. IEE Intelligent Systems Engineering, Winter 1993, pag.246-256.

[42] Bristol E. H. On a new measure of interaction for multivariable process control. IEEE Transactions 1966, vol. AC-11, pp. 133-134.

[43] Fararooy S., Perkins J. D., Malik T. I., Oglesby M. J., Williams S. Process controllability toolbox (PCTB). Computers Chem. Engng. 1993, vol. 17, no. 5/6, pp. 617-625.

[44] Kwatny H., Maffezzoni C. Control of electric power generating plants. In: Levine W S (ed), The Control Handbook, CRC Press - IEEE Press, 1996, pp. 1453-1483.

[45] Bolis V., Maffezzoni C., Ferrarini L. Synthesis of the overall boiler-turbine control system by single loop auto-tuning technique. IFAC J. Control Engineering Practice, 3, 6, pp. 761-771, 1995.

[46] Isidori A. Nonlinear Control Systems. 3rd edn. Springer, London, 1995.

[47] Passino K. M., Yorkovich S. Fuzzy control. In: Levine W S (ed), The Control Handbook, CRC Press - IEEE Press, 1996, pp. 1001-1017.

[48] Farrel J. A. Neural control. In: Levine W S (ed), The Control Handbook, CRC Press - IEEE Press, 1996, pp. 1017-1030.

[49] Isermann R. On the applicability of model based fault-detection for technical processes. Control Engineering Practice 1994, vol. 2, no. 2, pp. 439-450.

[50] Morari M., Zafirion E. Robust Process Control. Prentice Hall Int., 1989.

[51] Sahlin P., Bring A., Sowell E. The neutral model format for building simulation. Swedish Inst. of Applied Mathematics, ITM Report no. 1992:3, Göteborg, March 1992.

[52] Bartolini A., Leva A., Maffezzoni C., Power Plant Simulator Embedded in a Visual Programming Environment, Proc. IFAC Symp. SIPOWER '95, Cancún 1995, pp. 119-124.

[53] Simon D., Espian B., Castillo E., Kapellos K. Computer-aided design of a generic robot controller handling reactivity and real-time control issues. IEEE Trans. on control Systems Technology 1993, vol. 1, no. 4, pp. 213-229.

[54] Otter M., Grübel G. Direct physical modeling and automatic code generation for mechatronics simulation. Proc. 2nd Conf. on Mechatronics and Robotics, Duisburg, 1993.

[55] Blankenship G. L., Ghanadau R., Kwatny H. G., La Vigna C., Polyakov V. Tools for integrated modeling, design and nonlinear control. IEEE Control Systems 1995, vol. 15, no. 2, pp. 65-77.

[56] SpeedUp^{TM}. Aspen technology. 1991.

[57] Rutz R., Richert J. CAMeL: an open CACSD environment. IEEE Control Systems 1995, vol. 15, no. 2, pp. 26-33.

[58] Ferretti G., Magnani G., Putz P., Rocco P. The structural design of an industrial robot controller. Control Engineering Practice 1996, vol. 4, no. 2, pp. 239-249.

[59] Lausterer G. K. Economic benefits of advanced control of power plants. Proc. of International BIAS Conf., Milan (Italy), 1993.

[60] Lausterer G. K. The preview computer: a new aid for supervision and control of complex systems. Proc. of INTERKAMA Congress 92, Düsseldorf (Germany), 1992, pp. 106-115.

[61] Ramming J., Wagner R. Optimization computer reduces costs for fuel and external power supplies. Energy and Automation 1990, vol. 12, pp. 43-44.

[62] Lausterer G. K., Klinger-Reinhold R., Seibert E. A knowledge-based operator system - Concepts, knowledge acquisition and practical experience. I MECH E Seminar on the application of expert systems in the power generation industry, London, Nov. 17, 1993.

[63] Richalet J. Industrial applications of model based predictive control. Automatica, Vol. 29, No. 5, pp. 1251-1274, 1993.

6. Advances in Variable Structure Control

Claudio Bonivento*   Roberto Zanasi*

Abstract

Variable Structure Control is a switching feedback control which
provides a simple tool for coping with uncertain nonlinear plants.
The computer technology and high-speed switching circuitry have
made the implementation of VSC of increasing interest to control
engineers. The control action is used in order to maintain the system state trajectory on a prescribed sliding surface. However,
because of non-ideality of switching the problem of chattering
arises, which constitutes a major limitation in real-world applications. Recently, research efforts have been made in order to
offer modifications and extensions of the basic theory to alleviate
this problem. The lecture, collecting recent results by the authors,
some of which are already available separately in the literature, fo-
cuses on the Discontinuous Integral Control (DIC) method both in
the continuous and discrete-time case. Some application examples
both to simulated and real systems illustrate the methodological
approach.

1   Introduction

The problem of robust stabilization of uncertain systems has been studied and solved by using several control methods. Among these, the

*Dipartimento di Elettronica Informatica e Sistemistica (DEIS), Università di
Bologna, Viale Risorgimento, 2, 40136 Bologna, Italia, Tel.: +39-51-6443045
(Bonivento), -6443034 (Zanasi); Fax: +39-51-6443073; http://www-lar.deis.unibo.it/;
E-mail: {cbonivento,rzanasi}@deis.unibo.it.

Variable Structure Control (VSC) approach [1], [2], [3] allows a total rejection of external disturbances under proper matching conditions. The VSC achieves robustness properties with respect to external and parametric disturbances by using discontinuous control variables. The main feature of VSC method is the sliding mode, which occurs when system motion is along the designed switching surfaces.

Since the first appearance of this theory, several VSC algorithms have been proposed in the literature. The interested reader can refer to survey papers such as [4], [5], which contain a very extensive list of references. The main idea of these algorithms is to use a feedforward action to compensate the known part of the system, and then introduce a feedback switching term in the control variable in order to eliminate the undesired unknown disturbances. Since almost all these algorithms are *not dynamic*, in order to remain in the sliding mode the amplitude of the switching term must be larger than the amplitude of the disturbance.

In many practical situations, because of the presence of non-idealities in the controlled plant, the implementation of a VS regulator does not guarantee the *ideal* sliding mode, but only that the states reach and stay within some boundary layer of the sliding surface. In these cases the system outputs are affected by high-frequency oscillations whose amplitude is usually proportional to the amplitude of the control switching term. This is the so-called chattering phenomenon. Often, these high-frequency oscillations lead to undesirable behaviours of the closed loop system. A natural way to reduce the influence of these oscillations on the system is to try to decrease their amplitude by decreasing the amplitude of the switching term in the control law as much as possible.

The chattering attenuation is the main issue addressed in this lecture. More precisely, some VSC algorithms forcing the closed loop system to remain in the sliding mode when the amplitude of the switching term is smaller than the amplitude of the disturbance are here presented and discussed. The main feature of these control algorithms is the fact that they are *dynamic*: integrators with nonlinear switching inputs are introduced with the aim of tracking the unknown disturbances. The estimates given by the integrators are used in the control algorithm to partially compensate the disturbances. The difference between the real disturbances and the available estimates is then eliminated by using a small-amplitude switching term. This is the basic and simple idea underlying the method, which is indicated as Discontinuous Integral Control (DIC).

While VSC theory was first introduced in the continuous-time case, more investigations on the discrete-time case are necessary for a satisfactory implementation in computer-controlled systems (see for example [6], [7], [8]). In this context, we introduce a few discrete-time DIC algorithms, for which the concept of *discrete-time sliding mode* is used for referring to system trajectories forced to stay in a proper neighborhood of a given manifold.

The DIC design originally formulated for the continuous-time case [9], [10], [11], [12], [13] is discretized and properly modified in order to introduce a derivative action on the controlled variable. The so-obtained discrete-time controller proves to be very effective in the disturbance estimation and in the chattering reduction, especially when the system parameters are known and constant. If this is not the case, two further adaptive control structures are considered to cope with the uncertainty on the upper bound of the disturbance variation [14] and on the inertia parameter of the system [15].

Since the basic purpose of this lecture is to offer a general application-oriented overview of a class of results about the topic of chattering reduction, most of the stability and robustness proofs are not reported here. For such details, the interested reader can refer to the indicated references. On the other hand, in order to give a feeling of the DIC applicability to different technical areas, the last significant part of the lecture is dedicated to the illustration of a few application examples with some remarks on the obtained experimental results.

More precisely, the lecture is organized as follows. In Section 2 the basic control problem and the classical VSC solution are recalled. Section 3 contains the central results of the DIC method for the continuous-time case. Moreover, a relaxed version with saturated control is provided along with a theoretical interpretation in terms of the Binary Control [16], [17], [18], [19]. The DIC discrete-time case is addressed in Section 4. Firstly, a simple discretized version of the algorithm is considered, for which the previously mentioned improved version with derivative action is also discussed. Secondly, a further discrete-time algorithm which adapts the amplitude k of the switching term to the estimate of the disturbance variation is introduced. Finally, an adaptive generalization is given for the case when the "inertia" of the system is unknown. Section 5 reports on three application cases referring to the regulation of a CST chemical reactor [20], [13], the velocity control of an electric DC motor driving a highly time-varying mechanical load [21], [22], and the point-to-point position control of a robotic system [23], [24].

2   Basic problem and classical solution

Let us consider the following uncertain system:

$$\dot{x} = \tilde{A}(x,t)x + \tilde{B}(x,t)u + \overline{\varphi}(x,t) + \overline{\psi}(t) \qquad (1)$$

where $x \in R^n$ is the state, $u \in R^m$ is the control, $\tilde{A}(x,t)$ and $\tilde{B}(x,t)$ are the time-varying matrices of the system, and $\overline{\varphi}(x,t)$ and $\overline{\psi}(t)$ are the parametric and external disturbances, respectively. It is always possible to express matrices $\tilde{A}(x,t)$ and $\tilde{B}(x,t)$ as follows

$$\begin{cases} \tilde{A}(x,t) &= A + \overline{A}(x,t) \\ \tilde{B}(x,t) &= B + \overline{B}(x,t) \end{cases} \qquad (2)$$

where the constant matrices A and B are the known part of the system, and the time-varying matrices $\overline{A}(x,t)$ and $\overline{B}(x,t)$ the uncertain one. Using (2), system (1) becomes:

$$\dot{x} = Ax + Bu + D(x,t) \qquad (3)$$

where

$$D(x,t) = \overline{A}(x,t)x + \overline{B}(x,t)u + \overline{\varphi}(x,t) + \overline{\psi}(t)$$

is the global disturbance acting on the system.

Assumption 1. *Suppose that disturbance $D(x,t)$ acts only in the range space of B: $\mathcal{R}(D(x,t)) \subseteq \mathcal{R}(B)$. From this assumption ("matching condition"), it follows that there exist two functions $\varphi'(x,t)$ and $\psi(t)$ such that*

$$D(x,t) = B[\varphi'(x,t) + \psi(t)] \qquad (4)$$

Therefore, system (1) becomes

$$\dot{x} = Ax + B[u + \varphi'(x,t) + \psi(t)] \qquad (5)$$

Uncertain functions $\psi(t) : R_+ \to R^m$ and $\varphi'(x,t) : R^n \times R_+ \to R^m$ are supposed to satisfy the conditions of existence and uniqueness of the solutions of (5).

Assumption 2. *Suppose that pair* (A, B) *is controllable and that*

$$rank(B) = rank(CB) = m \tag{6}$$

where $\sigma_x = Cx = 0$ $(C \in R^{(m \times n)})$ *is the relation between the state variables (so-called "sliding surface") expressing the desired dynamic behaviour of the controlled system.*

We desire now to transform system (5) in order that the first $n - m$ state coordinates belong to the sliding surface, and the last m coordinates belong to the subspace spanned by the input matrix B. To this purpose we can consider a two-steps procedure. First, applying the following state space transformation

$$y = T_B x \qquad \text{with} \qquad T_B = [E \,|\, B]^{-1}$$

where E is any matrix such that T_B is nonsingular, system (5) becomes

$$\begin{cases} \dot{y}_1 &= \hat{A}_{11}\, y_1 + \hat{A}_{12}\, y_2 \\ \dot{y}_2 &= \hat{A}_{21}\, y_1 + \hat{A}_{22}\, y_2 + u + \varphi''(y, t) + \psi(t). \end{cases} \tag{7}$$

where

$$y = \begin{bmatrix} y_1 \\ y_2 \end{bmatrix}, \qquad y_1 \in R^{n-m}, \qquad y_2 \in R^m$$

The sliding surface results as

$$\sigma_x = Cx = C T_B^{-1} y = C\, E y_1 + C\, B y_2 = 0 \tag{8}$$

From Assumption 2, we can left-multiply equation (8) by matrix $[C\,B]^{-1}$ and obtain

$$\sigma_y = [C\,B]^{-1} C\, E y_1 + y_2 = C_1 y_1 + y_2 = 0$$

where $C_1 = [C\,B]^{-1} C\, E \in R^{m \times (n-m)}$. As a second step, we define the state space coordinate transformation

$$T = \begin{bmatrix} I_{n-m} & 0 \\ C_1 & I_m \end{bmatrix}, \qquad T^{-1} = \begin{bmatrix} I_{n-m} & 0 \\ -C_1 & I_m \end{bmatrix}$$

where

$$\begin{bmatrix} z_1 \\ z_2 \end{bmatrix} = T \begin{bmatrix} y_1 \\ y_2 \end{bmatrix}, \qquad T \in R^{n \times n}, \qquad \det T = 1$$

so that $z_1 = y_1$ e $z_2 = \sigma_y = C_1 y_1 + y_2$. System (7) reduces now to the desired final form

$$\begin{cases} \dot{z}_1 = A_{11} z_1 + A_{12} z_2 \\ \dot{z}_2 = A_{21} z_1 + A_{22} z_2 + u + \varphi(z,t) + \psi(t) \end{cases} \qquad (9)$$

where in particular

$$A_{11} = \hat{A}_{11} - \hat{A}_{12} C_1$$

Note that from the controllability assumption on (A, B) and then on $(\hat{A}_{11}, \hat{A}_{12})$, it follows that by properly choosing matrix C_1 it is possible to define arbitrarily the eigenvalues of matrix A_{11}. Since the dynamic behaviour of subsystem (i.e. the sliding mode dynamics)

$$\dot{z}_1 = A_{11} z_1 \qquad (10)$$

can be arbitrarily assigned by transformation T, the aim of the feedback design is to force $z_2 = 0$ for the closed loop system, in spite of the presence of the external disturbance. This problem can be solved using the VSC approach [1] and [2]. The state z of system (9) will reach the sliding mode surface $z_2 = 0$ with an exponential λ_0-decay of state vector z_2, if the following sufficient conditions are satisfied

$$\text{Sgn}\,(z_2)(\dot{z}_2 + \lambda z_2)|_{z_2 \neq 0} < \epsilon < 0 \qquad (11)$$

where $\text{Sgn}\,(z_2) = diag[\,\text{sgn}\,z_{21}, \ldots, \text{sgn}\,z_{2m}]$ and $\lambda = diag[\lambda_1, \ldots, \lambda_m]$ with $\lambda_i \geq \lambda_0$ for $i = 1, \ldots, m$. Usually, in the classical VSC solution, the external disturbance is supposed to satisfy the following assumptions.

Assumption 3. *For $\forall t \in R_+$ the uncertain function*

$$\varphi(z,t) = [\varphi_1, \ldots, \varphi_m]^T$$

satisfies the conditions

$$|\varphi_i(z,t)| < \phi_i(z), \qquad \phi_i(0) = 0, \quad \dot{\phi}_i(0) < \infty \qquad (12)$$

for $i = 1, \ldots, m$, where $\Phi = diag[\phi_1, \ldots, \phi_m]$ is a known vector function $\forall x \in R^n$.

Assumption 4. *For $\forall t \in R_+$ the uncertain function*

$$\psi(t) = [\psi_1, \ldots, \psi_m]^T$$

satisfies the conditions

$$|\psi_i(t)| < \Delta_{0i} \qquad (13)$$

for $i = 1, \ldots, m$, where $\Delta_0 = diag[\Delta_{01}, \ldots, \Delta_{0m}]$ is a known constant vector.

Classical Solution. The control law

$$u = \underbrace{-A_{21}z_1 - A_{22}z_2}_{u_1} \quad \underbrace{-\lambda z_2}_{u_2} \quad \underbrace{-(\Phi + \Delta_0)\operatorname{sgn} z_2}_{u_3} \qquad (14)$$

where $\operatorname{sgn} z_2 = [\operatorname{sgn} z_{21}, \ldots, \operatorname{sgn} z_{2m}]^T$, satisfies conditions (11). Therefore, after a finite time $t' > 0$ a sliding mode arises on surface $z_2 = 0$, and the dynamic behaviour of closed loop system (9)-(14) is described by equation (10). In (14), control variable u consists of three terms corresponding to three different control actions: u_1 compensates for the undesired reaction $A_{21}z_1 + A_{22}z_2$ of the known part of system (9), u_2 gives the proportional action that ensures the exponential decay λ_0 of state vector z_2, and u_3 is the discontinuous control action that instantaneously compensates the external disturbance. When sliding mode is reached, u_2 is zero, u_1 tends exponentially to zero and control law u tends to u_3.

Since u_3 is a *static* function of state z and time t, it can be effective in the complete rejection of the disturbance if and only if the amplitude $\Phi + \Delta_0$ of the switching term is larger than or equal to the amplitude of disturbance $\varphi + \psi$. This means that, using control law (14), when the external disturbance is large, the switching term must also be large.

From a theoretical point of view, when the system is in the sliding mode the amplitude value of the switching term does not affect the dynamic behaviour (10) of the closed loop system. For example, the amplitude of u_3 can be twice or ten times the minimum needed value $\Phi + \Delta_0$ without any particular problem. From a practical point of view the situation can be different. In fact, if small non-idealities are considered (i.e. delays, unmodeled dynamics, etc.), the use of VS feedback (14) does not ensure the ideal sliding mode and the system output are affected by high-frequency oscillations (*chattering*) whose amplitude is almost proportional to the amplitude Δ_0 of switching term u_3. So, a reduction of the amplitude of u_3 means, in practical cases, also a reduction of the amplitude of the chattering oscillations.

If the only information available on external disturbance $\varphi + \psi$ are Assumptions 3 and 4, then the only VSC law ensuring that the system remains in the sliding mode on surface $z_2 = 0$ is control law (14), and Δ_0 is the minimum possible amplitude of switching term u_3.

When more information is available on the external disturbance, in some cases it is possible to guarantee the sliding mode on $z_2 = 0$ also with a smaller amplitude of u_3. This argument introduces the concept of Discontinuous Integral Control illustrated in the following sections.

3 Discontinuous Integral Control: continuous-time case

The main idea underlying the following developments is the introduction of a simple dynamic mechanism for estimating the system disturbance in order to reduce in a substantial way the amplitude of the switching term of the classical solution.

Discontinuous Integral Control (DIC) solution. The general DIC structure is the following:

$$\begin{cases} u = -A_{21}z_1 - (A_{22} + \lambda)z_2 - (\Phi + k)\operatorname{sgn} z_2 - \tilde{\psi} \\ d\tilde{\psi}/dt = h \operatorname{sgn} z_2 \end{cases} \tag{15}$$

where $k = diag[k_1, \ldots, k_m]$, $h = diag[h_1, \ldots, h_m]$, $\lambda = diag[\lambda_1, \ldots, \lambda_m]$ are constant vectors with $k_i > 0$, $h_i > 0$ and $\lambda_i > 0$ for $i = 1, \ldots, m$, and $\tilde{\psi} = diag[\tilde{\psi}_1, \ldots, \tilde{\psi}_m]$ are the disturbance estimators.

The innovations to the classical solution (14) are:

a) Coefficients k_i can be smaller than the amplitudes Δ_{0i} of disturbance $\psi(t)$ (see Assumption 4).

b) In (15), m integrators are introduced. The aim of these integrators $\tilde{\psi}_i$ is to estimate the components $\psi_i(t)$ of the disturbance acting on the system. These estimations $\tilde{\psi}_i$ are used in the controller as a feed-forward actions that partially compensate disturbance $\psi(t)$ allowing a reduction of the switching term u_3.

Introducing the new state variable

$$z_3 = \psi - \tilde{\psi} \tag{16}$$

representing the error between disturbance ψ and estimate $\tilde{\psi}$, the equations of closed loop system (9), (15) become:

$$\begin{cases} \dot{z}_1 &= A_{11}z_1 + A_{12}z_2 \\ \dot{z}_2 &= \varphi - \lambda z_2 - (\Phi + k)\,\mathrm{sgn}\,z_2 + z_3 \\ \dot{z}_3 &= \dot{\psi} - h\,\mathrm{sgn}\,z_2 \end{cases} \qquad (17)$$

Since the second and third equations of system (17) are nearly in diagonal form and are function of state variable z_1 only through disturbance $\varphi(z,t)$, the analysis of the $2m$-order subsystem $\{z_2, z_3\}$ is equivalent to the analysis of m 2-order decoupled subsystems $\{z_{2i}, z_{3i}\}$ for $i = 1, \ldots, m$. In fact, if control law (15) guarantees $z_{2i}(t) = 0$ in a finite time t_i^* for each subsystem $\{z_{2i}, z_{3i}\}$, then for $t > max\{t_i^*\}$ the external disturbance will be completely rejected and the dynamic behaviour of the closed loop system will be the desired one (10). For this reason, without losing generality, in the following we will analyze only the case $m = 1$, that is $z_2, z_3 \in R$.

The detailed proofs of the results reported in this section are in [11].

Assumption 5. *The disturbance $\psi(t)$ has a bounded derivative and bounded initial condition:*

$$|\dot{\psi}(t)| \le \Delta_1, \qquad |\psi(0)| < \Delta_0 \qquad (18)$$

with $\Delta_1 > 0$ and $\Delta_0 > 0$ known constants.

Then, the following results hold.

Result 1. *Let consider control law (15) when $\varphi(z,t) = 0$, $\lambda = 0$, $z_2(0) = 0$ and disturbance $\psi(t)$ satisfies Assumption 5. If for arbitrary $k > 0$, coefficient h satisfies the inequality*

$$h > \frac{\Delta_1}{2}\left(\frac{\Delta_0}{k} + \frac{k}{\Delta_0}\right) \qquad (19)$$

then the controlled system is globally stable, and after a finite time t^* a sliding mode arises on surface $z_2 = 0$. $\square$

The stability analysis reduces to the analysis of the following second order system:

$$\begin{cases} \dot{z}_2 &= z_3 - k\,\mathrm{sgn}\,z_2 \\ \dot{z}_3 &= \dot{\psi} - h\,\mathrm{sgn}\,z_2 \end{cases} \qquad (20)$$

with initial conditions $[z_2(0)| = 0, z_3(0)]$, and in presence of the *worst disturbance*

$$\dot{\psi}_w = \Delta_1 \, \text{sgn} \, \dot{z}_2 \tag{21}$$

In Fig. 1 the typical behaviour of the system trajectories in the phase plane (z_2, z_3) is reported.

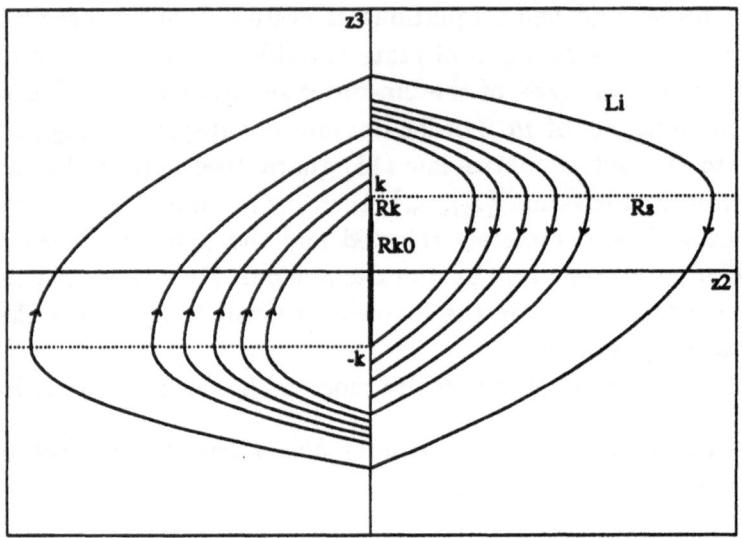

Figure 1: Typical (z_2, z_3) phase-portrait for the worst disturbance case.

Region $R_k := \{z_2 = 0, |z_3| < k\}$ is the only region of the state plane in which a sliding mode is present. Once state z reaches region R_k, it cannot exit from it. In the phase plane, it exists a stability region R_s, see Fig. 3, delimited by the limit cycle L_i, such that if the initial condition $z(0)$ belongs to R_s, the system is stable, while if $z(0) \notin R_s$ there exist disturbance functions $\psi(t)$ satisfying assumption (18) for which the system is unstable.

Remark 1. Result 1 implies that given initial conditions $z_2(0) = 0$, $z_3(0)$, and an arbitrarily positive value for k, it is always possible to choose a proper value for h such that the controlled system is stable. When the system reaches the sliding mode condition on $z_2 = 0$, estimation error z_3 is bounded $(|\psi - \tilde{\psi}| \leq k)$ and the amplitude of the switching term tends exponentially to k. Parameter k can be chosen arbitrarily small, but it cannot be zero $(k = 0)$ because in this case amplitude $\bar{z}_{30}$ of the stability region in the worst case is zero.

Remark 2. In Fig. 2 it is illustrated as limit cycle L_i is deformed in these two cases: a) k is constant and h is decreased linearly; b) k is decreased linearly and h is constant. It can be noted that when $k \to 0$ the area of stability region R_s tends to zero, while when $h \to \Delta_1$ the area of region R_s remains finite.

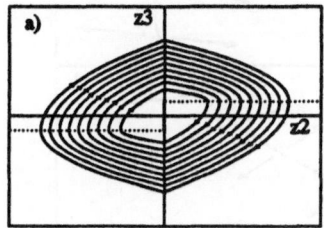
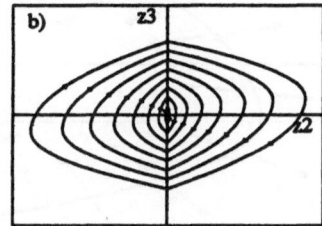

Figure 2: Changes in the limit cycle when: a) k is constant and h is decreased linearly; b) k is decreased linearly and h is constant.

If in control law (15) the proportional action is present, that is the parameter λ is not zero, the following result holds.

Result 2. *Let consider control law (15) with $\varphi(z, t) = 0$ and disturbance $\psi(t)$ with bounded derivative as in (18). If parameters $k_i > 0$, $h_i > \Delta_{1i}$ and $\lambda_i > 0$ $(i = 1, \ldots, m)$ satisfy the following inequalities*

$$h_i \lambda_i k_i > \Delta_{1i}^2 (1 - \ln 2) \tag{22}$$

then system (9) is globally stable and in finite time the sliding mode on surface $z_2 = 0$ is reached. □

When $\lambda > 0$ and $\varphi(z, t) = 0$, the analysis of the controlled system (17) reduces, in analogy with the previous case (20), to the analysis of the following second order system

$$\begin{cases} \dot{z}_2 &= z_3 - \lambda z_2 - k \operatorname{sgn} z_2 \\ \dot{z}_3 &= \dot{\psi} - h \operatorname{sgn} z_2 \end{cases} \tag{23}$$

Because of the presence of proportional term λz_2 in control law (15), two limit cycles L_i and L_e can arise in the phase plane (z_2, z_3). These two limit cycles divide the phase plane in three regions: R_s, R_{ie} and R_e (see Fig. 4). If system state z is within region R_s or region R_e, the corresponding trajectory tends to go *towards* the origin; if $z \in R_{ie}$, the corresponding trajectory, in the worst case, tends to go *away* from the origin. When the value of λ increases, unstable region R_{ie} tends to

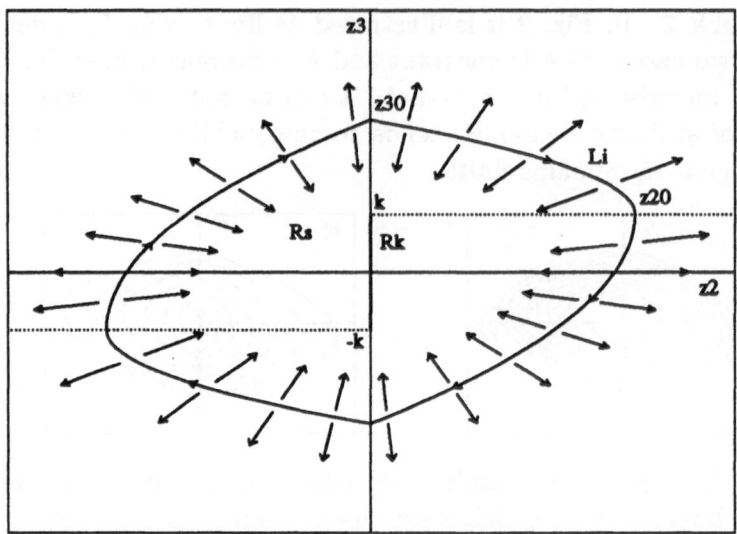

Figure 3: Stability region R_s delimited by limit cycle L_i.

reduce and for a particular value of $\lambda = \lambda^*$ it happens that $L_i = L_e$ and region R_{ie} disappears. Inequality (22) does not give the exact value for λ^*, but it gives a suitable approximation of it. It can be proved that if inequality (22) holds, for any initial condition $[z_2(0), z_3(0)]$ the system trajectory, in finite time, enters sliding mode region R_k and thereafter remains within it.

Furthermore, it can be shown [11] that the presence of parametric disturbance $\varphi(z,t)$ does not affect substantially the stable behaviour of the controlled system: it just modifies the dynamic behaviour during the transient.

It can be interesting for applications to consider the case where the ideal switching is substituted by a saturation characteristic. To this topic is dedicated the following section.

3.1   The case of saturated control

Let us now consider again the system (23) and let us study its dynamic behaviour when the switching term sgn z_2 is substituted by a saturation, i.e. $k \operatorname{sgn} z_2 \to k\,sat(\frac{z_2}{\alpha})$, where

$$sat\,x = \begin{cases} x & |x| \leq 1 \\ 1 & |x| > 1 \end{cases}$$

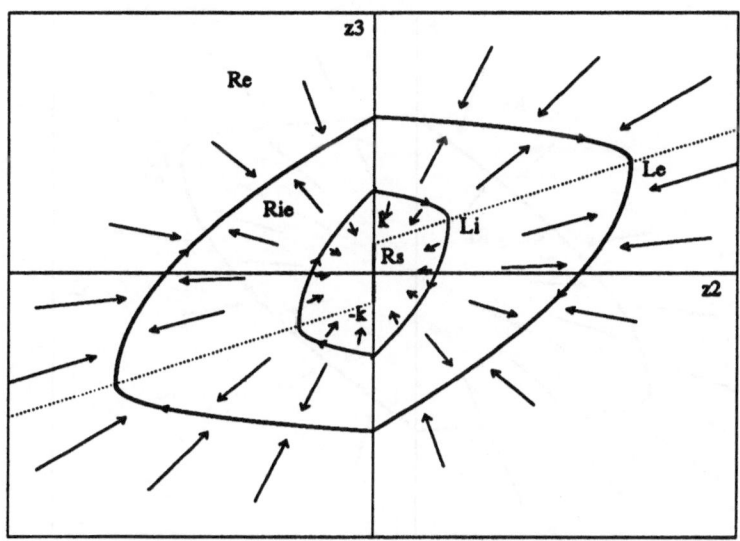

Figure 4: Limit cycles L_i and L_e in the phase plane when a small proportional term is used.

The parameter α represents the amplitude of the linear zone of the saturation. Let $\beta = \frac{k}{\alpha}$ denote the slope of the saturation function $k \, sat(\frac{z_2}{\alpha})$ in the linear zone, and let $\overline{\lambda} = \lambda + \beta$ be the global linear coefficient:

$$\beta = \frac{k}{\alpha}, \qquad\qquad \overline{\lambda} = \lambda + \beta$$

In the linear zone when $|z_2| < \alpha$, the system equations are:

$$\begin{cases} \dot{z}_2 &= -\overline{\lambda} z_2 + z_3 \\ \dot{z}_3 &= \dot{\psi} - h \operatorname{sgn} z_2 \end{cases} \qquad (24)$$

By integrating equations (24) in the presence of the *worst disturbance* (21), i.e. $\dot{\psi}_w = \Delta_1 \operatorname{sgn} \dot{z}_2$, and with initial conditions $z_2(0) = 0^+$ e $z_3(0) > 0$, one can obtain the following system trajectory in the (z_2, z_3) phase plane:

$$\overline{\lambda} z_2 = \left[\overline{\lambda} z_2(0) - z_3(0) - \frac{\overline{\Delta} \operatorname{sgn} z_2}{\overline{\lambda}} \right] e^{\frac{\overline{\lambda}[z_3 - z_3(0)]}{\overline{\Delta} \operatorname{sgn} z_2}} + z_3 + \frac{\overline{\Delta} \operatorname{sgn} z_2}{\overline{\lambda}} \qquad (25)$$

where $\overline{\Delta} = h - \Delta_1 \operatorname{sgn} \dot{z}_2$. As shown in Fig. 5, a stable limit cycle L_e is present in phase plane. This limit cycle constitutes the boundary of two

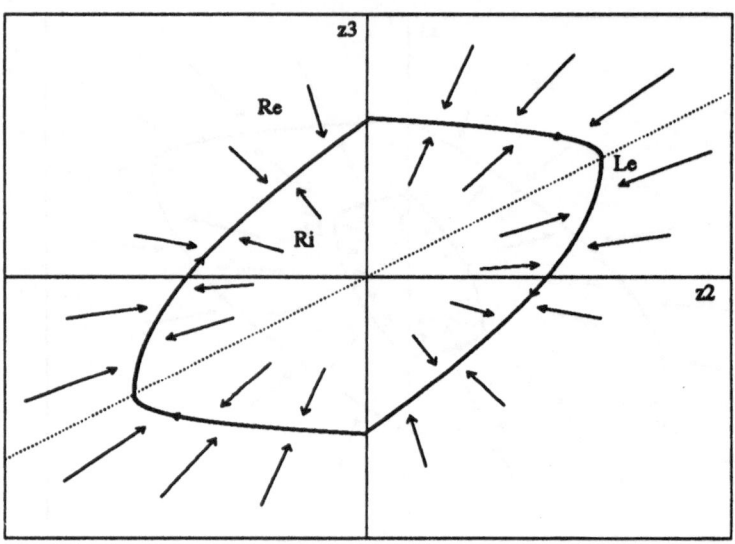

Figure 5: Limit cycle L_e in the phase plane.

regions: the external one R_e is stable, the internal one R_i is unstable. When the $\overline{\lambda}$ increases, the region R_i reduces but not vanishes in the limit. So, the control law (24) cannot guarantee in any case the *ideal* sliding mode.

Let $\bar{z}_{20}$ and $\bar{z}_{30}$ denote the maximum values of variables z_2 and z_3 along the limit cycle L_e. It is clear that a good control performance is achieved if the maximum value $\bar{z}_{20}$ is small enough with respect to the ideal case $z_2 = 0$. In this context, the following result proved in [13] is of interest because it provides an upper bound for $\bar{z}_{20}$ which explicitly depends on the design parameters h, λ and β.

Result 3. *In the worst case for the disturbance* $\dot{\psi} = \Delta_1\, sgn\, \dot{z}_2$, *system (24) has a limit cycle* L_e *for which*

$$\bar{z}_{30} < \frac{8\,\Delta_1}{5(\lambda + \beta)}$$

and

$$\bar{z}_{20} \leq \frac{64\Delta_1^2}{(25h + 15\Delta_1)(\lambda + \beta)^2}$$

□

The previous result is obtained under the constraint that the limit cycle is acting in the linear zone of the saturation function, i.e. $\bar{z}_{20} < \alpha$. From the obtained upper-bound result, it follows that the condition

$$\frac{64\Delta_1^2}{(25h + 15\Delta_1)(\lambda + \beta)^2} < \alpha$$

is sufficient to assure that the system trajectories enter and remain within the linear region.

3.2 Binary Control point of view

The basic concepts underlying the previously illustrated results can be seen also in the frame of the so-called Binary Control [16].

The goal of the binary control theory is the synthesis of a continuous non-linear control law u such that, from any given initial condition, the state trajectories $x(t)$ move towards the origin remaining in the vicinity of hyperplane $\sigma_x = 0$. The notion of *vicinity* is formalized by means of the following *hypercone*:

$$G_{\delta_0} = \{x \in R^n : |\sigma_x| \leq \delta_0 \|x\|\} \tag{26}$$

where δ_0 is a design parameter that defines the exact shape of the hypercone and $\|x\| = \sum_{i=1}^n |x_i|$. In the case of a second order system, the hyperplane $\sigma_x = 0$ is a straight line and the hypercone G_{δ_0} is a conic region of the plane inside which the trajectories are forced to move towards the origin.

The general control structure of the Binary Control is shown in Fig. 6. In this scheme, the variables associated with double arrows (as μ, ρ and ν) are *operator* variables, while the variables associated with single arrows are *coordinate* ones. A variable is referred to as *coordinate* if it is subjected to some transformation, or it is referred to as *operator* if it dictates the form of the transformation applied to some coordinate variable. This dual interpretation of the state variables of a non linear system is referred to as the *binarity principle*. The introduction of the operator variables proves to be useful, from a methodological point of view, during structural analysis of the systems.

In Fig. 6, P is the plant to be controlled, $y_d(t)$ is the reference signal, $\varphi(z,t)$ is an unknown parametric disturbance, R_u, R_μ, R_ρ and R_ν are operators defining four different control loops with specific properties:

R_u acts directly on the state coordinates of the system and defines a feedback of *coordinate* type;

R_μ changes the properties of R_u using the information on both the state variables and the desired dynamics $\sigma_x = 0$. It is called a feedback of *operator* type;

R_ρ through δ_0 and μ slightly relaxes the desired dynamics ($\sigma_x \to \overline{\sigma} = \sigma_x - \delta_0\mu(R_u \circ x)$), so that the time-varying surface $\overline{\sigma} = 0$ is reached with a global smoother system behaviour. This is the so-called Operator Feedback (OF);

R_ν (Operator Coordinate Feedback: OCF) defines the external control loop in order to guarantee that the closed loop system behaviour is asymptotically independent from the uncertain parameters of system P.

The general expression of the feedback law is the following:

$$\begin{cases} u & = & R_\mu \circ R_u \circ x + R_\nu \circ (x, \overline{\sigma}) \\ \overline{\sigma} & = & R_\rho \circ (x, \sigma_x) \end{cases} \tag{27}$$

where $A \circ B$ means that the operator A affects the behaviour of B. With different choices for R_u, R_μ, R_ρ and R_ν, it is possible to obtain well-known control schemes, such as VS, Bang-Bang, Adaptive, and so on, or design new ones. In particular, a basic structure showing good robustness properties when the external reference signal is zero ($y_d = 0$) is the following:

$$\begin{cases} u & = & R_\mu \cdot R_u \circ x + R_\nu \circ (x, \overline{\sigma}) \\ R_u \circ x & = & \overline{k}^T |x| \\ R_\mu & = & \mu(t) \\ R_\nu \circ (x, \overline{\sigma}) & = & -\eta \, sat\frac{\overline{\sigma}}{\|x\|} \\ \\ \overline{\sigma} & = & \sigma_x - \delta_0\mu(R_u \circ x) \\ \\ \dot{\mu} = \begin{cases} -\gamma \, \text{sgn} \, \overline{\sigma} & |\mu| \leq 1 \\ -\omega\mu & |\mu| > 1 \end{cases} \end{cases} \tag{28}$$

where $|\mu(0)| \leq 1$, $\overline{k} = [k_1, k_2, \ldots, k_n]^T$, $|\mathbf{x}| = [|x_1|, |x_2|, \ldots, |x_n|]^T$ is the vector of the absolute values of state variables, γ and η are positive constants chosen on the basis of known parameters of system P, ω is an arbitrary positive constant, $\sigma_x = 0$ is the desired dynamics, and δ_0 defines an hyperconic region G_{δ_0} in the state space containing the relaxed desired dynamics $\overline{\sigma} = 0$.

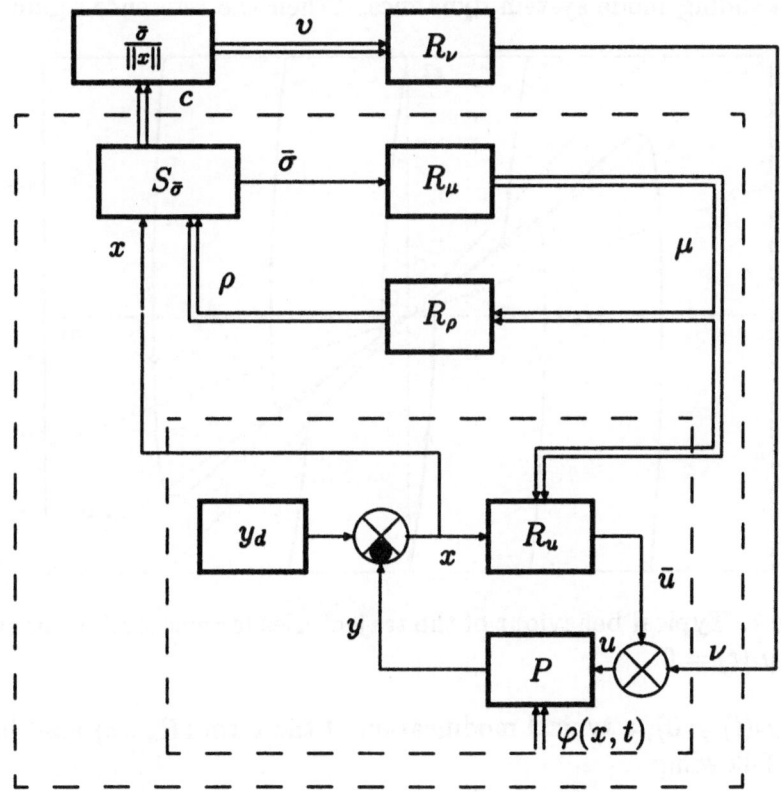

Figure 6: General Binary Control structure.

In Fig. 7 it is reported a typical behaviour of the state-space trajectories for a second order system with $y_d(t) = 0$ and control law (28). If parameters $\overline{k}$, γ, and η satisfy a set of proper inequalities given in [16], the trajectories of the closed loop system are forced to reach in a finite time and remain thereafter within the hypercone G_{δ_0}. This hypercone constitutes a global attractor and its amplitude δ_0 is inversely proportional to the γ control parameter. When $\gamma \to \infty$ the hypercone degenerates to an hyperplane, and the control becomes of the VS type with a sliding mode system dynamics. When the reference signal is not

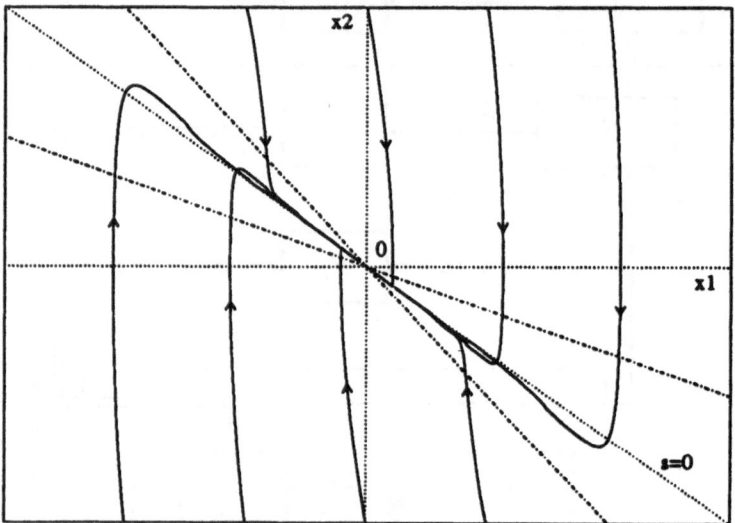

Figure 7: Typical behaviour of the trajectories for a second order system when $y_d(t) = 0$.

zero $(y_d(t) \neq 0)$, a typical modification of the term $(R_u \circ x)$ used in (28) is the following:

$$R_u \circ x = \overline{k}^T |x| + k, \qquad k > |\psi(t)| \qquad (29)$$

where $\psi(t)$ is a function of the exogenous disturbances (reference signal, external loads, and so on) and their time-derivatives acting on the system. Using control law (29), the system P can be stabilized with a steady-state error approaching zero only if the parameter γ approaches infinity. Again, in this limit case the Binary Control structure becomes a VS one, and the control signal u becomes discontinuous.

It is interesting to note that the DIC control structure can be seen both as a modification of the VS classical solution (14) or as a particular case of the binary control structure (28) by redefining the R_ν operator as follows:

$$\begin{cases} R_\nu \circ (x, \overline{\sigma}) = \nu(t) \\ \\ \dot{\nu} = \begin{cases} -h \operatorname{sgn} \overline{\sigma} & |\nu| \leq \nu_{max} \\ -\omega\nu & |\nu| > \nu_{max} \end{cases} \end{cases} \qquad (30)$$

and assuming $\gamma \to \infty$.

4   Discontinuous Integral Control: discrete-time case

As mentioned in the introduction, the definition of a discrete-time version of the DIC control law is of practical interest in the applications and implies some extensions of the previous results.

Let us consider again the control law (15). If

- the compensation of the known part of the system is achieved,
- the proportional action is disregarded
- and the parametric disturbance is absent $(\varphi(z, t) = 0)$,

then the formulation of the control law reduces to the following basic scheme

$$\begin{cases} u = -k \operatorname{sgn} y - \tilde{\psi} \\ d\tilde{\psi}/dt = h \operatorname{sgn} y \end{cases}$$

where now for notational simplicity we use y for z_2.

Direct discrete version of DIC law. Consider the following control loop equations:

$$\begin{cases} \dot{y}(t) & = & \psi(t) - k \operatorname{sgn} y(n) - \tilde{\psi}(n) \\ \tilde{\psi}(n) & = & \tilde{\psi}(n-1) + h\,T_s \operatorname{sgn} y(n) \end{cases} \qquad (31)$$

In the first equation of (31), a mixed notation has been used: $\dot{y}(t)$ and $\psi(t)$ are continuous-time variables, while $y(n)$ and $\tilde{\psi}(n)$ are piecewise-constant variables whose value changes only at sampling instants $t = n\,T_s$.

An important difference in the discrete control law (31) with respect to the continuous one is that estimation $\tilde{\psi}(n)$ is now sampled, that is, at each sampling time the increment of estimation $\tilde{\psi}(n)$ is equal to $\pm h\,T_s$ even if the disturbance is zero or constant. Consequently, an oscillation similar to the one produced by the switching term $k \operatorname{sgn} y$ is present on the controlled output. If we want to track a disturbance ψ with a large derivative $(\Delta_1 > |\dot{\psi}|)$, parameter h must be larger than Δ_1 $(h > \Delta_1)$, and therefore also the basic oscillation generated by the estimator $(h\,T_s)$ will be large. This oscillation is the main drawback of the discrete control law (31) with respect to the continuous one. Note that when $T_s \to 0$, the oscillation $h\,T_s$ also tends to zero $(\forall h)$.

In order to reduce the oscillation generated by the estimator, we introduce a proper modification of the discrete control law (31). By integrating the first equation in (31) in the time interval $[n\,T_s,\ (n+1)\,T_s]$ we obtain:

$$y(n+1) = y(n) - k\,T_s \operatorname{sgn} y(n) + \int_{n T_s}^{(n+1) T_s} \psi(t)\,dt - \tilde{\psi}(n)\,T_s \qquad (32)$$

Equation (32) describes the dynamic behaviour of the output y at the sampling instants $t = k\,T_s$. Let us denote with $\overline{\psi}(n+1)$ the mean value of the external disturbance $\psi(t)$ in the time interval $[n\,T_s,\ (n+1)\,T_s]$:

$$\overline{\psi}(n+1) = \frac{1}{T_s} \int_{n T_s}^{(n+1) T_s} \psi(t)\,dt$$

Equation (32) can now be written as follows:

$$\frac{y(n+1) - y(n)}{T_s} + k \operatorname{sgn} y(n) = \overline{\psi}(n+1) - \tilde{\psi}(n) \qquad (33)$$

The term on the left-hand side of equation (33) has the physical meaning of *mean estimation error* in the sampling interval $[n\,T_s,\ (n+1)\,T_s]$ and therefore seems to be a candidate function to be used in the disturbance estimator in place of the function $\operatorname{sgn} y$. With this modification the controlled system (31) becomes as follows.

Improved discrete version of DIC law. Let us consider

$$\begin{cases} \dot{y}(t) = \psi(t) - k \operatorname{sgn} y(n) - \tilde{\psi}(n) \\[2mm] \tilde{\psi}(n) = \tilde{\psi}(n-1) + k_e \left[\dfrac{\Delta y(n)}{T_s} + k \operatorname{sgn} y(n-1) \right] \end{cases} \qquad (34)$$

where $\Delta y(n) = y(n) - y(n-1)$ and k_e is a proper design parameter. The term $\Delta y(n)/T_s$ is the discrete time-derivative of output y generated in the last time interval, see (33), by the switching term $-k \operatorname{sgn} y(n-1)$ and the estimation error $\overline{\psi}(n) - \psi(n-1)$. Note that when the estimation error is zero, the estimator $\tilde{\psi}(n)$ is constant and therefore it does not cause any oscillation in the controlled system.

Let us now analyze system (34) at the sampling instants. In this case we have the following discrete equations:

$$\begin{cases} y(n+1) = y(n) - k\,T_s\, \text{sgn}\, y(n) + T_s \bar{\psi}(n+1) - \tilde{\psi}(n)\,T_s \\[2mm] \tilde{\psi}(n) = \tilde{\psi}(n-1) + k_e \left[\dfrac{\Delta y(n)}{T_s} + k\,\text{sgn}\, y(n-1) \right] \end{cases} \tag{35}$$

If we use the mean estimation error $e(n)$

$$e(n) = \frac{y(n) - y(n-1)}{T_s} + k\,\text{sgn}\, y(n-1) \tag{36}$$

as the new state variable, system (35) is transformed as follows

$$\begin{cases} e(n+1) & = & \bar{\psi}(n+1) - \tilde{\psi}(n) \\[2mm] \tilde{\psi}(n) & = & \tilde{\psi}(n-1) + k_e\, e(n) \end{cases} \tag{37}$$

In this form the system is linear with respect to the state variables $e(n)$ and $\tilde{\psi}(n)$. The *mean disturbance* $\bar{\psi}(n)$ can be considered as the new disturbance input. From (36) it follows that the dynamics of the output is completely determined by the estimation error $e(n)$:

$$y(n+1) = y(n) - k\,T_s\,\text{sgn}\, y(n) + e(n+1)\,T_s \tag{38}$$

For system (38), we have a discrete sliding mode if the amplitude of the switching term $k\,T_s$ is greater than the amplitude of disturbance $e(n)\,T_s$, that is, if

$$k > |e(n)|_{max} \tag{39}$$

If the estimation error $e(n)$ is smaller than k, the output y is forced to move towards the switching surface $y = 0$ and stay within the region

$$|y| < (k + |e|_{max})\,T_s, \qquad\qquad (k > 0) \tag{40}$$

The maximum value of the estimation error $e(n)$ depends on the disturbance $\bar{\psi}$ acting on the system. Let us now analyze system (37). From the first equation of the system it follows that

$$e(n+1) - e(n) = \bar{\psi}(n+1) - \bar{\psi}(n) - [\tilde{\psi}(n) - \tilde{\psi}(n-1)] \tag{41}$$

By using the second equation of system (37) $(\tilde{\psi}(n) - \tilde{\psi}(n-1) = k_e\, e(n))$, we finally obtain

$$e(n+1) + (k_e - 1)\, e(n) = \overline{\psi}(n+1) - \overline{\psi}(n) \tag{42}$$

This difference equation describes the dynamics of the estimation error as a function of the disturbance $\overline{\psi}(n)$. By using the $\mathcal{Z}$-transform, the linear equation (42) is described by the following discrete transfer function

$$\frac{e(z)}{\overline{\psi}(z)} = \frac{z-1}{z + k_e - 1} \tag{43}$$

The linear system (42) is stable if and only if the pole $p = 1 - k_e$ of the corresponding discrete transfer function is inside the unitary circle, that is

$$0 < k_e < 2 \tag{44}$$

One of the best choices for k_e when the disturbance is constant is $k_e = 1$, because in this case the pole p is in the origin (deadbeat condition). In the following we will consider only admissible values for k_e according to (44). Let us now analyze the dynamic behaviour of system (42) in the following cases when the external disturbance $\psi(t)$ is: i) a step function; ii) a ramp; iii) a sine function; iv) or when disturbance $\overline{\psi}(n)$ has a bounded discrete derivative.

i) **Step disturbance.** If the external disturbance $\psi(t)$ is a step function ($\psi(t) = c$), the disturbance $\overline{\psi}(t)$ is equal to the same step function delayed by one period T_s:

$$\psi(z) = \frac{cz}{z-1} \qquad \leftrightarrow \qquad \overline{\psi}(z) = \frac{c}{z-1} \tag{45}$$

From (43), it follows that

$$e(z) = \frac{z-1}{z + k_e - 1}\, \overline{\psi}(z) = \frac{c}{z + k_e - 1} \tag{46}$$

By applying the final value theorem it results that in steady-state condition the estimation error $e(\infty)$ is always zero ($0 < k_e < 2$):

$$e(\infty) = \lim_{z \to 1} \frac{z-1}{z} \frac{c}{z + k_e - 1} = 0 \tag{47}$$

This means that the control structure (35) is able to exactly estimate the external disturbance $\psi(t)$ if it is constant.

ii) **Ramp disturbance.** If the external disturbance $\psi(t)$ is a ramp with slope v ($\psi(t) = v\,t$), the disturbance $\overline{\psi}(t)$ has the following expression

$$
\begin{aligned}
\overline{\psi}(n) &= \frac{1}{T_s} \int_{(n-1)T_s}^{nT_s} v\,t\,dt = \frac{v}{2\,T_s} \left[t^2\right]_{(n-1)T_s}^{nT_s} \\
&= \frac{v}{2\,T_s} \left[n^2\,T_s^2 - (n-1)^2 T_s^2\right]
\end{aligned}
\tag{48}
$$

By applying the $\mathcal{Z}$-transformation we obtain

$$
\begin{aligned}
\overline{\psi}(z) &= \frac{v}{2\,T_s} \left[T_s^2 \frac{z\,(z+1)}{2\,(z-1)^3} - T_s^2 \frac{(z+1)}{2\,(z-1)^3}\right] \\
&= \frac{v\,T_s}{2} \frac{z+1}{2\,(z-1)^2}
\end{aligned}
\tag{49}
$$

and therefore, from (43), the estimation error $e(t)$ is:

$$
e(z) = \frac{v\,T_s}{2} \frac{z+1}{(z-1)(z+k_e-1)}
\tag{50}
$$

The steady-state value $e(\infty)$ of the estimation error is constant:

$$
e(\infty) = \lim_{z \to 1} \frac{z-1}{z} \frac{z+1}{(z+k_e-1)(z-1)} \frac{v\,T_s}{2} = \frac{v\,T_s}{k_e}
\tag{51}
$$

So, the estimator of control structure (35) is able to follow an external ramp disturbance $\psi(t)$, showing a constant tracking error which is proportional to slope v and inversely proportional to parameter k_e. From (39) we know that control structure (35) remains in the discrete sliding mode ($y = 0$) if the amplitude k of the switching term is larger than the estimation error, that is, if:

$$
k > \frac{|v|\,T_s}{k_e}
\tag{52}
$$

Only ramps can be tracked: it can be shown that if the disturbance is a parabolic function, the estimation error goes to infinity.

iii) **Sinusoidal disturbance.** Let us consider a sinusoidal external disturbance $\psi(t) = \sin \omega t$. Disturbance $\overline{\psi}(n)$ is the following:

$$
\begin{aligned}
\overline{\psi}(n) &= \frac{1}{T_s} \int_{(n-1)T_s}^{nT_s} \sin \omega t\,dt \\
&= \frac{1}{\omega\,T_s}(\cos \omega\,(n-1)\,T_s - \cos \omega\,n\,T_s)
\end{aligned}
\tag{53}
$$

The $\mathcal{Z}$-transform $\overline{\psi}(z)$ is:

$$\overline{\psi}(z) = \frac{1}{\omega T_s} \frac{(1-z)(z - \cos \omega T_s)}{z^2 - 2z \cos \omega T_s + 1} \tag{54}$$

The estimation error is found to be the following:

$$e(z) = \frac{z-1}{z+k_e-1} \frac{1}{\omega T_s} \frac{(1-z)(z - \cos \omega T_s)}{z^2 - 2z \cos \omega T_s + 1} \tag{55}$$

In this case the final value theorem cannot be applied because the sequence $e(n)$ has no limit for $n \to \infty$. As will be seen, in the steady-state condition the estimation error $e(n)$ becomes a sinusoidal oscillation: we are interested in finding the amplitude of this oscillation. Function $e(z)$ in (55) can be written as the product of two terms:

$$e(z) = F(z) \mathcal{Z}[\cos \omega t] \tag{56}$$

where $\mathcal{Z}[\cos \omega t]$ is the $\mathcal{Z}$-transform of $\cos \omega t$

$$\mathcal{Z}[\cos \omega t] = \frac{z(z - \cos \omega T_s)}{z^2 - 2z \cos \omega T_s + 1}$$

and $F(z)$ is the discrete transfer function of a proper linear system

$$F(z) = \frac{1}{\omega T_s} \frac{(z-1)^2}{z(z+k_e-1)}$$

Since the estimation error $e(z)$ can be interpreted as the output of the linear system $F(z)$ when the input is a sinusoidal function, then the amplitude of $e(z)$ in the steady-state condition ($|e|_\infty$) is equal to $|F(z)|$ when $z = e^{j\omega T_s}$: $|e|_\infty = |F(z)|_{z=e^{j\omega T_s}}$. It can be shown that

$$|e|_\infty = \frac{2(1 - \cos \omega T_s)}{\omega T_s \sqrt{k_e^2 - 2(k_e-1)(1 - \cos \omega T_s)}} \tag{57}$$

When $k_e = 1$ the estimation error reduces to $|e|_\infty = 2(1 - \cos \omega T_s)/\omega T_s$ and when $\omega T_s \to 0$ the estimation error tends to zero:

$$\lim_{\omega T_s \to 0} |F(e^{j\omega T_s})| = 0 \tag{58}$$

Let $\omega_s = 2\pi/T_s$ denote the sampling frequency. Fig. 8 shows the estimation error (57) as a function of ω/ω_s when $k_e = 1$. The maximum amplitude of the error ($|e|_\infty \simeq 1.5$) is reached for $\omega = 0.37 \omega_s$. From the

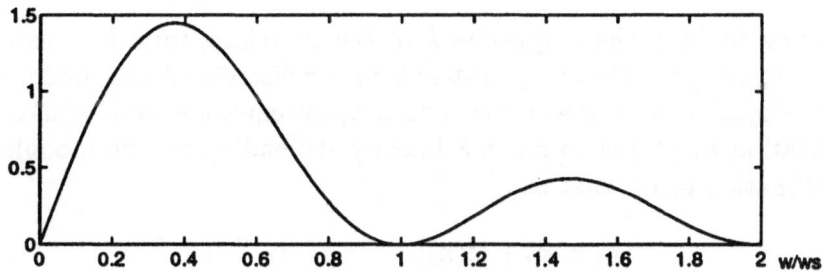

Figure 8: Estimation error $|e|_\infty$ as a function of ω/ω_s when $k_e = 1$.

figure it is evident that the estimation error is small when the sampling frequency ω_s is much higher than the disturbance frequency ω. For example, if $\omega_s > 20\omega$ then $|e|_\infty < 0.31$.

iv) Disturbance $\overline{\psi}(n)$ with bounded derivative. Let $\delta_\psi(n)$ denote the following variable

$$\delta_\psi(n) = \overline{\psi}(n) - \overline{\psi}(n-1) \tag{59}$$

which is proportional to the discrete derivative $\dot{\overline{\psi}}(n)$ of the disturbance $\overline{\psi}(n)$: $\delta_\psi(n) = \dot{\overline{\psi}}(n)T_s$. Equation (42) can now be rewritten as

$$e(n+1) + (k_e - 1)\,e(n) = \delta_\psi(n+1) \tag{60}$$

Since (60) is a first order linear stable system ($0 < k_e < 2$), when input $\delta_\psi(n)$ is bounded ($|\delta_\psi(n)| < \Delta_\psi$) the estimation error $e(n)$ is forced to stay, in the steady-state condition, within the bounded region $[-e_\Delta(\infty),\ e_\Delta(\infty)]$ where

$$e_\Delta(\infty) = \frac{\Delta_\psi}{k_e}$$

is the maximum steady-state error corresponding to the input $\delta_\psi(n) = \Delta_\psi$. The sliding mode condition (39) in this case becomes

$$k > \frac{\Delta_\psi}{k_e} = \frac{T_s|\dot{\overline{\psi}}(n)|_{max}}{k_e} \tag{61}$$

From (61) it follows that when $T_s \to 0$ the amplitude k of the switching term can be arbitrarily small, while for any finite $T_s > 0$ the amplitude k is lower-bounded.

4.1   Adaptive modification of the discontinuous gain

According to (34), the amplitude k of the switching term is constant, and therefore the control loop system is in the discrete sliding mode only if the estimation error $e(n)$ is lower then k, see equation (39). A natural modification in (34) is to make k linearly depending on the module of the estimation error, that is:

$$k(n) = k_0 + |e(n)|, \qquad\qquad k_0 > 0 \qquad\qquad (62)$$

Parameter k_0 is a small positive constant. When error $e(n)$ is zero, the estimation $\tilde{\psi}$ in (34) completely rejects the disturbance, and so the closed loop system can remain in the sliding mode even if the amplitude of the switching term is arbitrarily small $k = k_0$. On the contrary, if disturbance $\overline{\psi}$ has a bounded derivative, the estimation error is also bounded, see equation (61). In this case, according to equation (62) the value of $k(n)$ increases in order to keep the system in the sliding mode.

Adaptive discrete version of DIC law. The closed loop system equations related to the adaptive control law (62) are now the following:

$$\begin{cases} y(n+1) = y(n) - k(n)\,T_s\,\mathrm{sgn}\,y(n) + T_s\overline{\psi}(n+1) - \tilde{\psi}(n)\,T_s \\ \tilde{\psi}(n) \quad = \tilde{\psi}(n-1) + k_e\,e(n) \\ k(n) \quad = k_o + |e(n)| \\ e(n) \quad = \dfrac{y(n) - y(n-1)}{T_s} + k(n-1)\,\mathrm{sgn}\,y(n-1) \end{cases} \qquad (63)$$

As shown in [14], this algorithm implies that the maximum amplitude of the controlled variable y, when the system is in the sliding mode, is

$$|y| < (k_0 + 2\Delta_\psi)T_s \qquad\qquad (64)$$

where Δ_ψ is the upper-bound of the estimation error: $|e(n)| < \Delta_\psi$.

Let us now compare through simulations the direct discrete version of the DIC controller (31), the improved DIC structure (34) and the adaptive version (63). Firstly, the following parameters are used in (31) and (34):

$$T_s = 0.1\,s, \qquad k = 0.2, \qquad h = 1.5, \qquad k_e = 1$$

The disturbance $\psi(t)$ acting on both systems is a triangular wave with

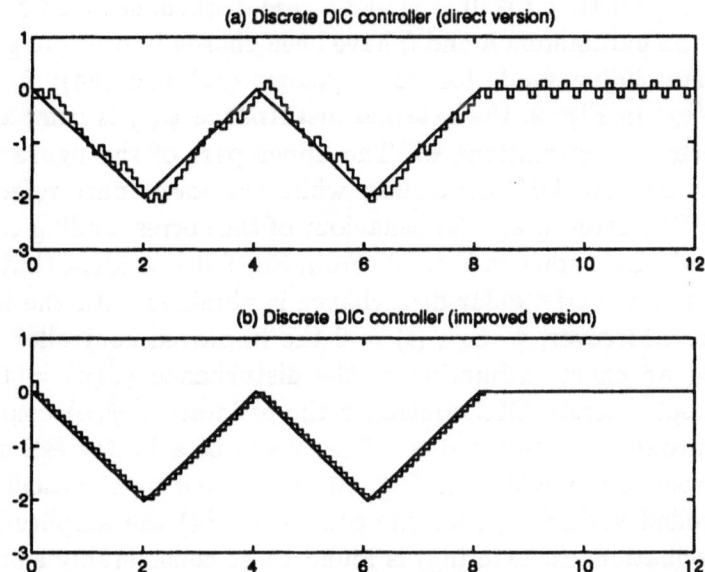

Figure 9: External disturbance $\psi(t)$ and disturbance estimations $\tilde{\psi}(n)$: a) Direct discrete version of DIC law (31) ; b) Improved discrete version of DIC law (34).

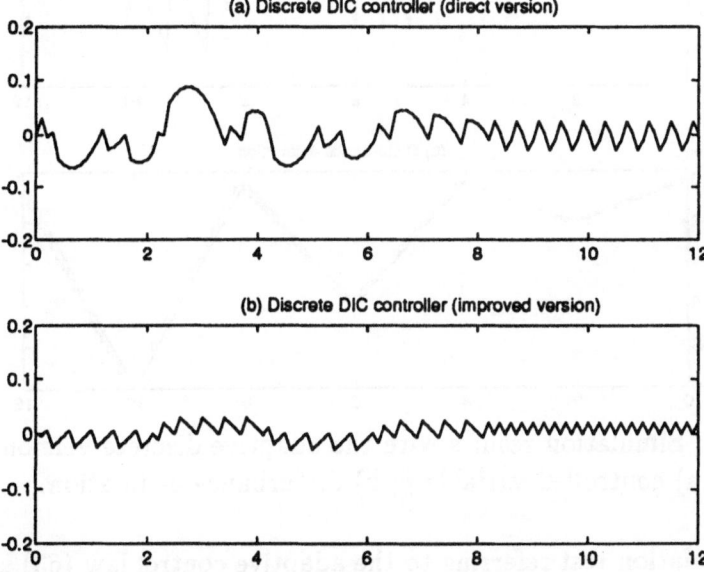

Figure 10: Behaviour of the controlled variables $y(t)$: a) Direct discrete version of DIC law (31) ; b) Improved discrete version of DIC law (34).

slope $v = \pm 1$ for time $t \in [0, 8\,s]$ and a zero disturbance for $t > 8\,s$. The values of the parameters h and k have been chosen in order to guarantee the discrete sliding mode for both systems (31) and (34): $h > v$ and $k > v\,T_s/k_e$. In Fig. 9, the external disturbance $\psi(t)$ is compared with the disturbance estimations $\tilde{\psi}$. The upper part of the figure refers to the direct discrete DIC controller, while the lower part refers to the improved DIC structure. The behaviour of the corresponding controlled variables $y(t)$ are shown in Fig. 10. From Fig. 9 it is evident that a better estimation of the triangular disturbance is obtained with the improved controller. Moreover, when $\psi(t) = 0$ the improved controller succeeds in making an exact estimation of the disturbance ($\tilde{\psi}(n) = 0$), while in the direct discrete DIC controller the estimation $\tilde{\psi}(n)$ continues to oscillate around the zero value. The effects of a better estimation of the disturbance are evident in Fig. 10: on comparing the oscillations of the controlled variables y, for the controller (34) the amplitude of the output oscillations (chattering) is found to be considerably smaller.

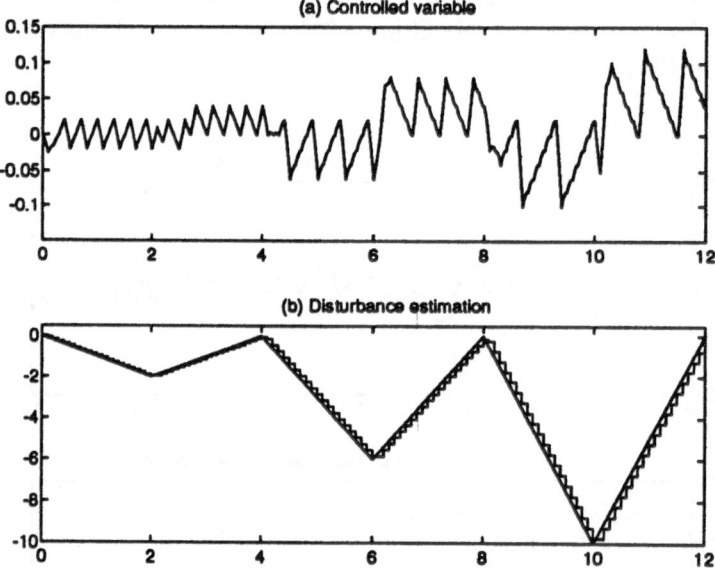

Figure 11: Simulation results with the adaptive discrete version of DIC law (63): a) controlled variable y; b) disturbance estimation.

A simulation test referring to the adaptive control law (63) is shown in Fig. 11. System parameters are: $T_s = 0.1\,s$, $k_0 = 0.2$ and $k_e = 1$. The triangular-wave disturbance assumes different slopes ($\pm 1, \pm 3, \pm 5$) for $t \in [0, 4\,s]$, $t \in [4, 8\,s]$ and $t \in [8, 12\,s]$, respectively. In the upper part of Fig. 11, the behaviour of the output y is shown, while the comparison of the external disturbance and its estimation is reported in the lower part. Note that the system keeps the discrete sliding-mode in spite of the increasing slope values of the disturbance. Without gain adaptivity (62), according to the tracking condition (52), the constant lower-bound on the gain k should be calculated for the maximum slope value, that is $k = 0.5$.

4.2 Estimation of the system "inertia"

As a further simple generalization let us now consider the following system

$$J\dot{y} = \psi(t) + u(t) \tag{65}$$

where J is a parameter (here called *inertia*) which is supposed to be constant but unknown or slowly time-varying. Let $J_e(n)$ denote the *estimation of the inertia J* at the sampling instant $t = nT_s$. The discrete controlled system is now the following

$$
\begin{cases}
y(n+1) & = & y(n) + \dfrac{T_s}{J}[-k \operatorname{sgn} y(n) + \bar{\psi}(n+1) - \tilde{\psi}(n)] \\[2mm]
\tilde{\psi}(n) & = & \tilde{\psi}(n-1) + \bar{e}(n) \\[2mm]
\bar{e}(n) & = & \dfrac{J_e(n-1)\,\Delta y(n)}{T_s} + k \operatorname{sgn} y(n-1)
\end{cases}
\tag{66}
$$

where $\bar{e}(n)$ can be considered as a *pseudo error* on the estimation of the disturbance. The *true error* $e(n)$ is:

$$e(n) = \frac{J\,\Delta y(n)}{T_s} + k \operatorname{sgn} y(n-1) \tag{67}$$

The pseudo error $\bar{e}(n)$ differs from true one $e(n)$ because it uses the inertia $J_e(n-1)$, estimated at the previous instant $t = (n-1)T_s$, instead of using the true inertia J. The errors $e(n)$ and $\bar{e}(n)$ are related by the equation

$$\frac{\bar{e}(n) - k \operatorname{sgn} y(n-1)}{e(n) - k \operatorname{sgn} y(n-1)} = \frac{J_e(n-1)}{J} \tag{68}$$

When $J_e \to J$, $\bar{e}(n) \to e(n)$. From (66) and (67) one obtains the dynamics of the estimation error $e(n)$:

$$
e(n+1) + \left[\frac{J_e(n-1)}{J} - 1\right] e(n) =
$$
$$
= \Delta\bar{\psi}(n+1) - k\left[1 - \frac{J_e(n-1)}{J}\right]\operatorname{sgn} y(n-1)
\tag{69}
$$

Note that when $J_e = J$ and $k_e = 1$, the equation (69) reduces to the equation (42). The homogeneous system corresponding to (69) is stable if and only if

$$\left| \frac{J_e(n-1)}{J} - 1 \right| < 1 \quad \longrightarrow \quad J_e(n-1) < 2J \qquad (70)$$

Inequalities (70) clearly show that when J is not known it is better to underestimate the initial value of J_e to ensure the stability of the controlled system. Due to the presence of the last term in the right-hand side of eq. (69), the estimation error $e(n)$ is not zero even if the disturbance $\overline{\psi}(n+1)$ is constant $(\Delta \overline{\psi}(n+1) = 0)$. To ensure in (66) both the asymptotic estimation of the true value J and the stability of the equilibrium point, let us consider the following updating equation

$$J_e(n) = J_e(n-1) + q\,\Delta\overline{e}(n) \frac{\Delta \operatorname{sgn} y(n-1)}{2} \qquad (71)$$

where $\Delta\overline{e}(n) = \overline{e}(n) - \overline{e}(n-1)$, $\Delta \operatorname{sgn} y(n-1) = \operatorname{sgn} y(n-1) - \operatorname{sgn} y(n-2)$ and $q > 0$ is a proper design parameter. In (71) the term $[\Delta \operatorname{sgn} y(n-1)]/2$ has the function to activate the updating mechanism only when the controlled system state is in the neighborhood of the sliding mode surface $y = 0$, that is when $y(n)$ changes the sign: $\operatorname{sgn} y(n-2) = -\operatorname{sgn} y(n-1)$. We will denote the control structure (66), (71) as *J-adaptive discrete DIC structure*. The following result holds.

Result 4. *Let us consider the controlled system (66)-(71) when the external disturbance $\psi(t)$ and the inertia J are constant. If the positive parameter q satisfies the inequality*

$$q < \frac{J}{2k} \qquad (72)$$

then the overall controlled system is asymptotically stable in a neighborhood of the equilibrium point $e(n) = 0$ and $J_e(n) = J$.
□

The proof of this result is reported in Appendix A.

When disturbance $\psi(t)$ and inertia J are constant and unknown, inequality (72) is a sufficient condition for the asymptotic estimation: $\tilde{\psi}(n) \to \psi$ and $J_e(n) \to J$. Since inertia J is supposed unknown, in the synthesis of the parameter q we have to use a lower-bound J_{min} of the inertia J, that is $q < J_{min}/(2k)$. Note that Result 1 can also be extended: when the disturbance $\psi(t)$, instead of being constant, is a ramp ($\Delta\bar{\psi} = \varphi_0$), it can be proved that the new equilibrium point $e(n) = \varphi_0$, $J_z(n) = 0$ is asymptotically stable when q is sufficiently small. When inertia J is slowly time-varying, system (86) does not guarantee the asymptotic estimation of $J(t)$, but still it shows good tracking performances.

As a simulation example, let us consider the block scheme of Fig. 12 representing the velocity control of a DC motor where J is the inertia of

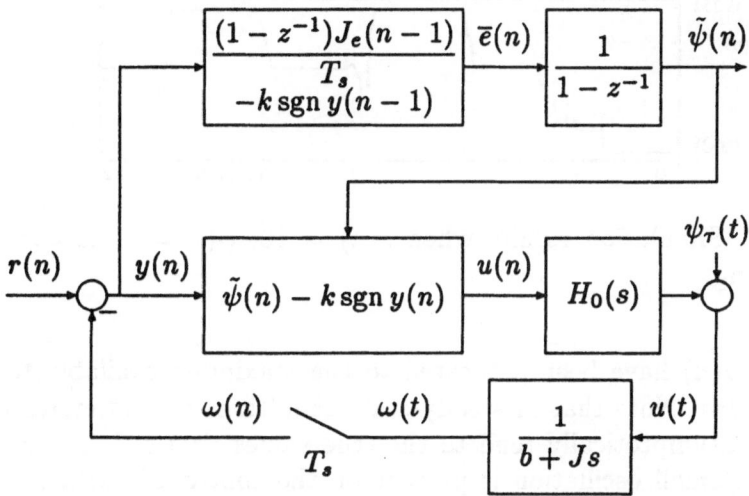

Figure 12: Block scheme used in simulation.

the motor, b the linear friction, $u(t)$ the input torque, $\psi_\tau(t)$ an external disturbance, $r(n)$ the reference signal, $y(n) = r(n) - \omega(n)$ the tracking error and $H_0(s)$ the zero order hold. The aim of the controller is to keep the tracking error $y(n)$ as small as possible even when the inertia J of the motor is time-varying. Fig. 13 reports the simulation results obtained when inertia J varies as follows

$$J(t) = 0.04 + 0.01 \operatorname{sgn}\left[\sin\left(15\,t\right)\right]$$

and when the disturbance is constant: $r(n) = 10$ and $\psi(t) = -7.5$. In the upper part of Fig. 13 the controlled variable $\omega(n)$ is reported together with the reference signal $r(n)$, while in the middle and low parts of the figure the estimated disturbance $\tilde{\psi}(n)$ and the estimated inertia $J_e(n)$ are shown. The used parameters are: $T_s = 0.005\,s$, $k = 0.8$, $q = 0.005$, $b = 0.002$ and $J_e(0) = 0.001$. The estimation $\tilde{\psi}(n)$ and

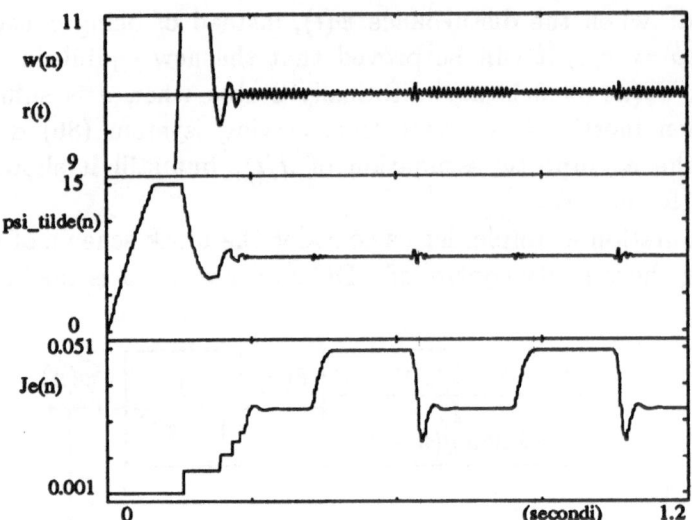

Figure 13: Simulation results when $r(n) = 10$, $\psi(t) = -7.5$ and J is time-varying.

the input $u(n)$ have been saturated to the maximum available torque $[-15, 15]Nm$. Note that in steady-state conditions the estimates $\tilde{\psi}(n)$ and $J_e(n)$ asymptotically tend to the true values $\psi(t)$ e J, despite the fact that a small oscillation is present on the controlled variable $\omega(t)$. Fig. 14 reports the results obtained when also the reference signal $r(n)$ and the disturbance $\psi_\tau(t)$ are time-varying:

$$r(n) = 10\cos(10\,nT_s), \qquad \psi_\tau(t) = -7\sin(8\,t)$$

The obtained simulation results show again the good performance of the *J-adaptive DIC structure*.

5   Application examples

In this section we illustrate the application of the DIC method in a variety of real world control problems. In particular we consider firstly the simulation study of the regulation of a chemical system. Then we present two fully implemented cases on physical systems: the velocity control of an electric DC motor and the point-to-point position control of a robotic system.

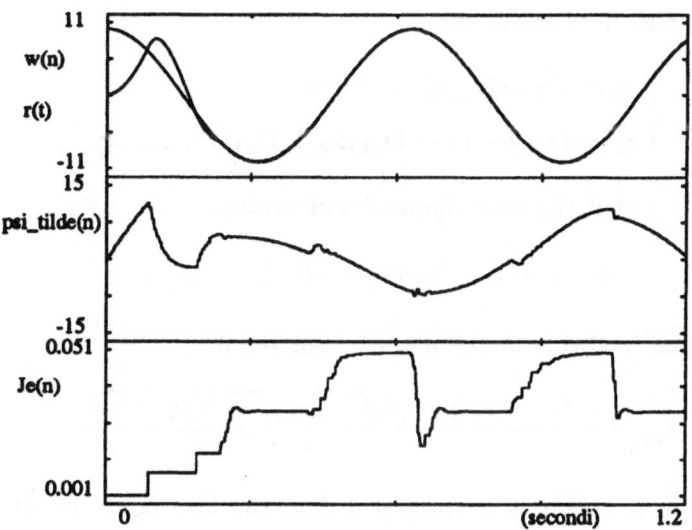

Figure 14: Simulation results obtained when J, $r(n)$ and $\psi(t)$ are time-varying.

5.1 Regulation of a Continuously Stirred Tank Reactor

We refer to the same system considered in [20]. The nonlinear dynamic model of a Continuously Stirred Tank Reactor (CSTR) in which an isothermal, liquid-phase, multi-component chemical reaction takes place is

$$\begin{cases} \dot{x}_1 &= -(1+D_{a1})x_1 + u \\ \dot{x}_2 &= D_{a1}x_1 - x_2 - D_{a2}x_2^2 \\ y &= x_1 + x_2 - C \end{cases} \tag{73}$$

where x_1 is the normalized concentration C_P/C_{P0} of a certain species P, with $C = C_{P0}$ being the desired total concentration of the species P and Q measured in mol.m^{-3}; x_2 is the normalized concentration C_Q/C_{P0} of species Q; u is the control variable, positive and upper bounded, defined as the ratio of the per-unit volumetric molar feed rate of species P and the desired total concentration C_{P0}. It is desired to regulate the total concentration error y toward zero, so that the total concentration value $x_1 + x_2$ converges to the specified set-point value C.

By using the following state variable transformation

$$\begin{cases} x_1 &= y - x_2 + C \\ x_2 &= x_2 \end{cases}$$

system (73) becomes as follows

$$\begin{cases} \dot{y} = -y - D_{a2}x_2^2 - C + u \\ \dot{x}_2 = D_{a1}y - (1 + D_{a1})x_2 + D_{a1}C - D_{a2}x_2^2 \end{cases} \tag{74}$$

We can note that if the *zero-dynamics* of system (74)

$$\dot{x}_2 = -(1 + D_{a1})x_2 + D_{a1}C - D_{a2}x_2^2 \tag{75}$$

is asymptotically stable in the equilibrium point

$$x_{2e} = \frac{-(1 + D_{a1}) + \sqrt{(1 + D_{a1})^2 + 4D_{a1}D_{a2}C}}{2D_{a2}}$$

the regulation problem is to force $y = 0$ in the first equation of system (74). In order to achieve this goal, let us consider the following DIC law

$$u = -\lambda y - k\,\mathrm{sgn}\,y - h \int \mathrm{sgn}\,y\,dt \tag{76}$$

By using the parameter values

$$D_{a1} = 1, \qquad D_{a2} = 1, \qquad C = 3$$

the equilibrium point is $x_1 = 2$, $x_2 = 1$. It is simple to verify that (75) is asymptotically stable. Using the controller parameter values

$$\lambda = 0.3, \qquad k = 0.05, \qquad h = 2$$

we obtain the time behaviours of state (x_1, x_2), input u and output y variables reported in Fig. 15.

Note that at times $t = 10\,s$ and $t = 15\,s$ parameter perturbations occur. In particular D_{a1} and D_{a2} take the value 1.5 on the time intervals $[10\,s, 11\,s]$ and $[15\,s, 16\,s]$, respectively. The simulation results obtained in the same operational conditions when the saturated control law analyzed in Section 3.1

$$u = -\lambda y - k\,\mathrm{sat}\,(\frac{\beta y}{k}) - h \int \mathrm{sgn}\,y\,dt \tag{77}$$

is used with $\beta = 10$, are reported in Fig. 16. In both cases, good properties are evident both for nominal regulation and for recovering, after disturbances, the desired stabilization features of the system.

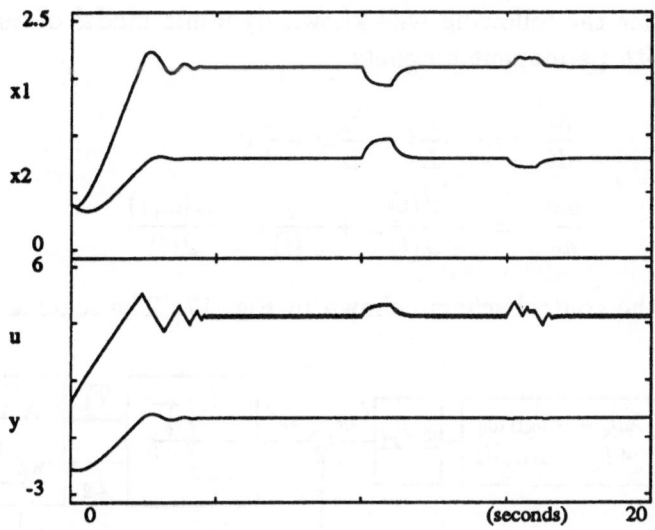

Figure 15: State variables x_1, x_2, input signal u and controlled variable y in the case of DIC control law (76).

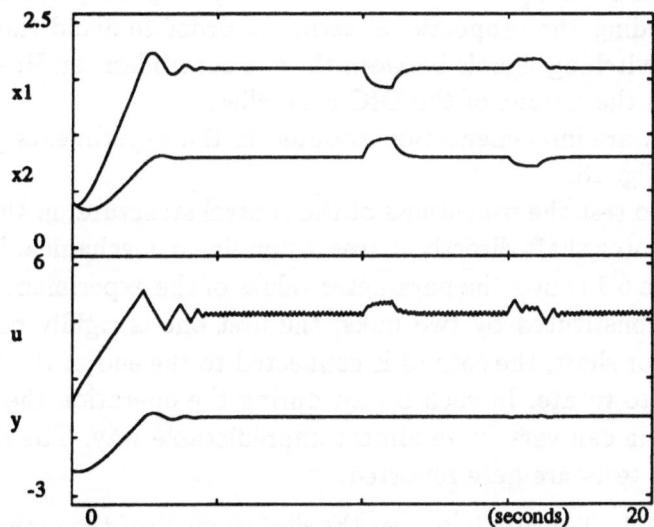

Figure 16: Variables x_1, x_2, u and y in the case of saturated control law.

5.2   Velocity control of an electric DC motor

Let us consider the following well-known dynamic model of an electric DC motor with permanent magnets:

$$\frac{di}{dt} = -\frac{r}{L}i - \frac{k_e}{L}\omega + \frac{1}{L}v \tag{78}$$

$$\frac{d\omega}{dt} = -\frac{\dot{J}(t)}{J(t)}\omega + \frac{k_c}{J(t)}i - \frac{c_r(\omega, t)}{J(t)} \tag{79}$$

We refer to the control scheme shown in Fig. 17. The scheme presents

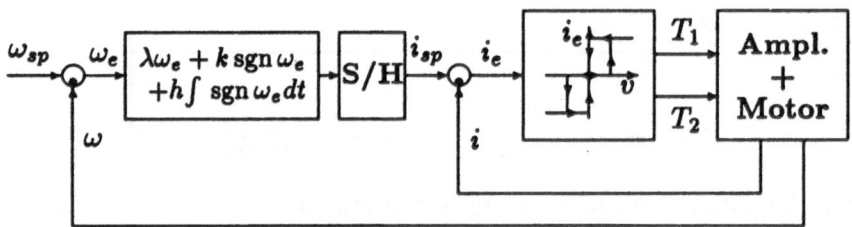

Figure 17: Velocity control scheme of the electric DC motor.

a cascade structure: in the inner current loop a typical VS control law with hysteresis is used, while the outer velocity loop adopts a DIC control law including the proportional term. In order to avoid the mutual influence of switching signals between the two controllers, an Hold circuit is inserted on the output of the DIC controller.

The hardware implementation adopted in the experiments [21], [22] is shown in Fig. 18.

In order to test the robustness of the control structure, in the experiments the motor shaft directly drives a non linear mechanical load, see Fig. 19. Table 6.1 shows the parameter values of the experimental setup. The load is constituted by two links, the first one is rigidly connected with the motor shaft, the second is connected to the end of the first link but it is free to rotate. In such a way, during the operation the value of the load inertia can vary in an almost unpredictable way. The results of four different tests are here reported.

Experiment 1. The aim is to test the performance of the inner current loop. To this purpose the velocity control loop is open, a sinusoidal current reference signal is applied with the swiching frequency of the power driver fixed to the value 6 kHz. The reference current amplitude is the

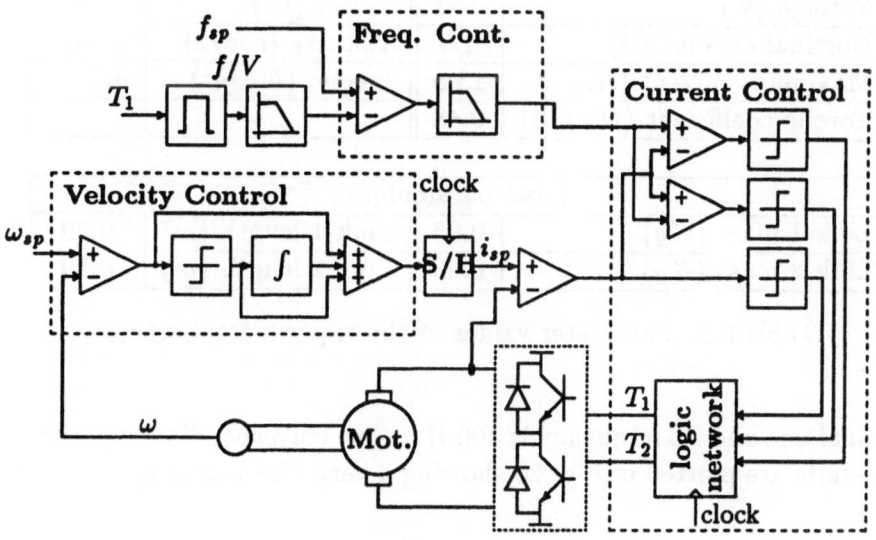

Figure 18: Hardware implementation adopted in the experiment.

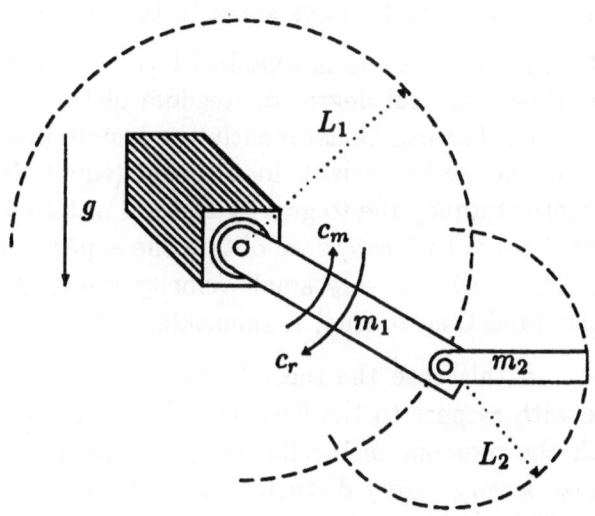

Figure 19: The mechanical load used in the experiments.

Motor parameters			
Voltage (V)	180	Power (W)	600
Nominal current (A)	4.5	Velocity (rad/s)	300
Maximum current (A)	13.5	Inertia $(Kg\,m^2)$	0.0011
Torque coefficient $(N\,m/A)$	0.48		

Load parameters			
Link 1 mass (Kg)	0.95	Link 1 length (m)	0.30
Link 2 mass (Kg)	0.29	Link 2 length (m)	0.10

Table 6.1: Parameter values of the experimental setup.

nominal one and its frequency is $300\,Hz$. The corresponding experimental results are plotted in Fig. 20 showing a very nice tracking behaviour.

Experiment 2. This experiment aims to test the dynamic performance of the velocity control loop when the mechanical load is disconnected. In this situation the system inertia and the residual load torque due to the friction are very small allowing fast responses. A trapezoidal velocity reference signal from 0 to 200 rad/s with acceleration value 4000 rad/s^2 is applied. The corresponding experimental results reported in Fig. 21 show a very low chettering and no overshoot in the tracking.

Experiment 3. In this case the mechanical load is connected to the motor shaft but the rotational degree of freedom of the second link is eliminated by a proper fixture. Consequently the system presents a constant moment of inertia and a variable load torque (equal about to 70% of the nominal motor torque) due to gravity effect. For a constant velocity reference signal equal to 6 rad/s we obtain the experimental results reported in Fig. 22 showing a very small velocity error. As expected, the motor current (and then torque) is sinusoidal in the mean.

Experiment 4. In this case the second link of the mechanical load is free to rotate with respect to the first link. Under these operational conditions, both the moment of inertia and the load torque are time varying and cause a very heavy disturbance condition on the system. For a constant velocity reference signal equal to 5 rad/s we obtain the experimental results reported in Fig. 23 still showing a very small velocity error despite a very large and irregular behaviour of the motor current.

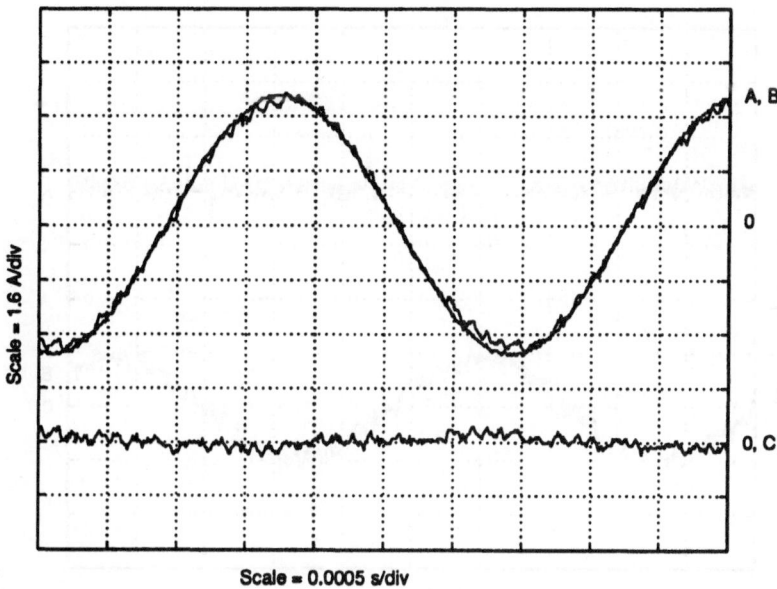

Figure 20: Experiment 1 results: A) reference current; B) actual current; C) current error.

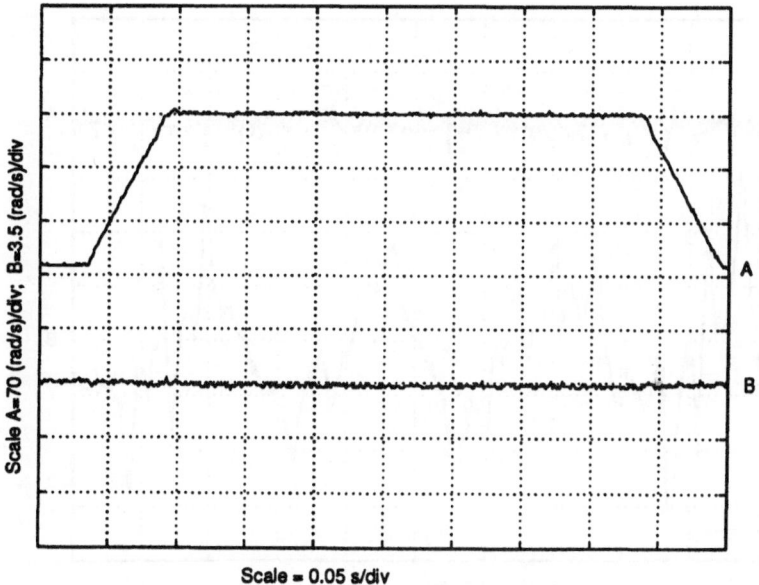

Figure 21: Experiment 2 results: A) actual motor velocity; B) velocity error.

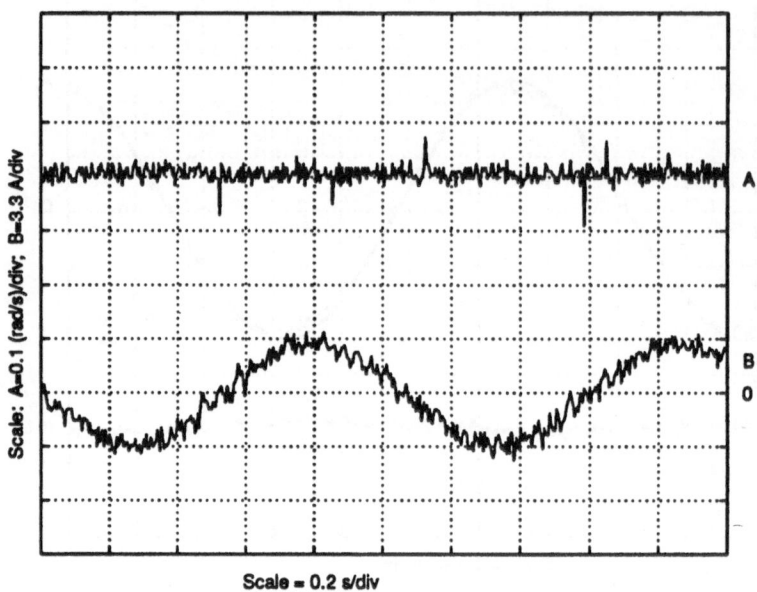

Figure 22: Experiment 3 results: A) velocity error; B) motor current.

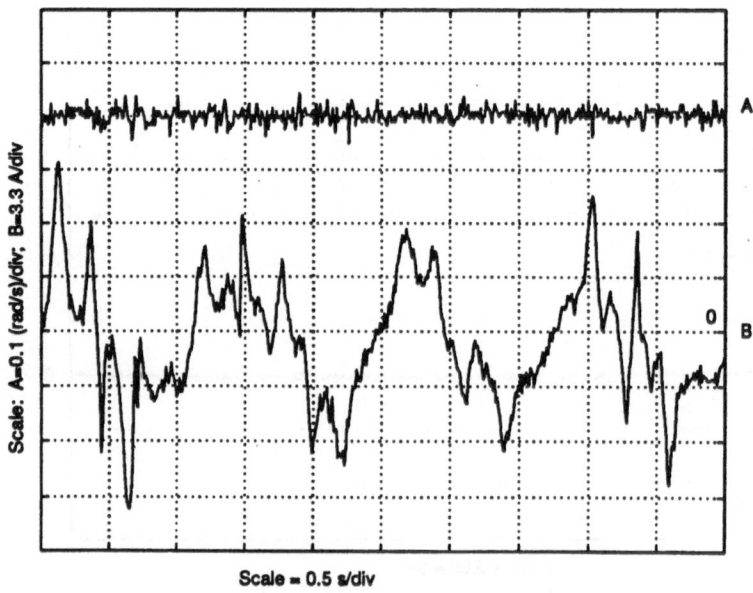

Figure 23: Experiment 4 results: A) velocity error; B) motor current.

5.3   Point-to-point position control of a robotic system

In this application we address a position control problem of the first joint of a industrial robot KUKA 361. More details about the system and the expertimental setup are in [23] and [24].

We describe the system as follows

$$M\ddot{q} = u(t) + \psi(t) \tag{80}$$

where M is the inertia of the system, $u(t)$ is the control variable, $\psi(t)$ is the total external disturbance acting on the system, and where $q = q_r - q_d$ is the position error between the actual position q_r and the desired position q_d. In the case of one-link mechanical system the total external disturbance is

$$\psi(t) = -f(\dot{q}_r) - g(q_r) - M\ddot{q}_d - \tau_e$$

where $f(\dot{q}_r)$ is the friction torque, $g(q_r)$ the gravitational torque and τ_e the load torque. For point-to-point movements we have that $\ddot{q}_d = \dot{q}_d = 0$ and therefore

$$q = q_r - q_d, \qquad\qquad \dot{q} = \dot{q}_r, \qquad\qquad \ddot{q} = \ddot{q}_r$$

The strategy for controlling to zero the state $(q, \dot{q})$ of system (80) is to keep the state on the following sliding mode surface

$$\sigma = \dot{q} + b(q) = 0 \qquad \text{where} \qquad b(q) = \frac{\dot{q}_m \omega_0 q}{\dot{q}_m + \omega_0|q|} \tag{81}$$

The shape of the adopted sliding mode surface $\sigma = \dot{q} + b(q) = 0$ in the phase plane $(q, \dot{q})$ is reported in Fig. 24. This choice automatically

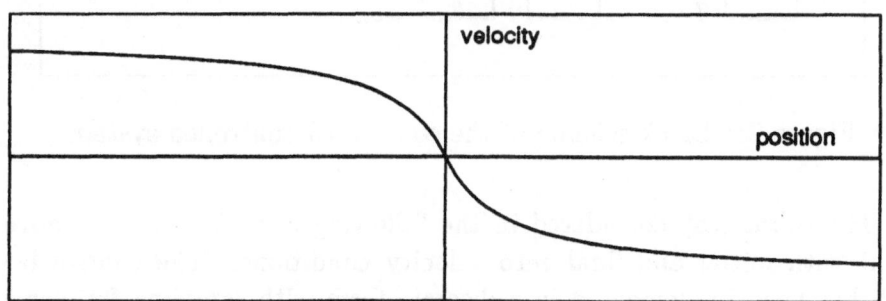

Figure 24: Sliding mode surface $\sigma = \dot{q} + b(q) = 0$ in the phase plane $(q, \dot{q})$.

imposes the upper bound $\dot{q}_m$ to the maximum allowed velocity during the transient and the slope value ω in the vicinity of the origin. An extended analysis on a class of candidate sliding mode surfaces for this problem can be found in [23].

In order to force the system state $(q, \dot{q})$ to reach and stay on the vicinity of the sliding mode surface $\sigma = 0$, we assume the following control law:

$$\begin{cases} u(t) = -K_0\sigma - r \\ \\ \dot{r} = \begin{cases} K_s \, \text{sgn} \, \sigma & \text{if} \quad |r| \le \gamma U_m \\ -\rho r & \text{if} \quad |r| > \gamma U_m \end{cases} \end{cases} \qquad (82)$$

where U_m is the maximum available torque, r is the integral action, K_0 and K_r are the linear and integral gains, and $0 < \gamma < 1$ and ρ are positive constants. One can note that in the integral action an anti-windup mechanism, i.e. an intrinsic limitation of the integrator output, has been introduced in order to cope with the physical saturation $[-U_m, U_m]$ in the system actuator.

The block scheme of the considered controlled system is shown in Fig. 25. In our application the inertia parameter value is $M = 40 \, Kg \, m^2$.

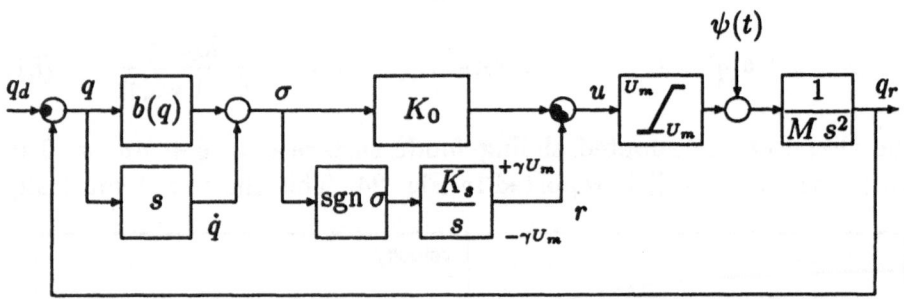

Figure 25: Block scheme of the considered controlled system.

The robot task considered in the following is a 45° rotation movement with initial and final zero velocity conditions. The control law (82) has been implemented in a discrete form with sampling frequency $f_s = 200 \, Hz$, using the trapezoidal approximation and introducing an estimate of the velocity from the direct measure of the incremental position values. The experimental results reported in Fig. 26 have been

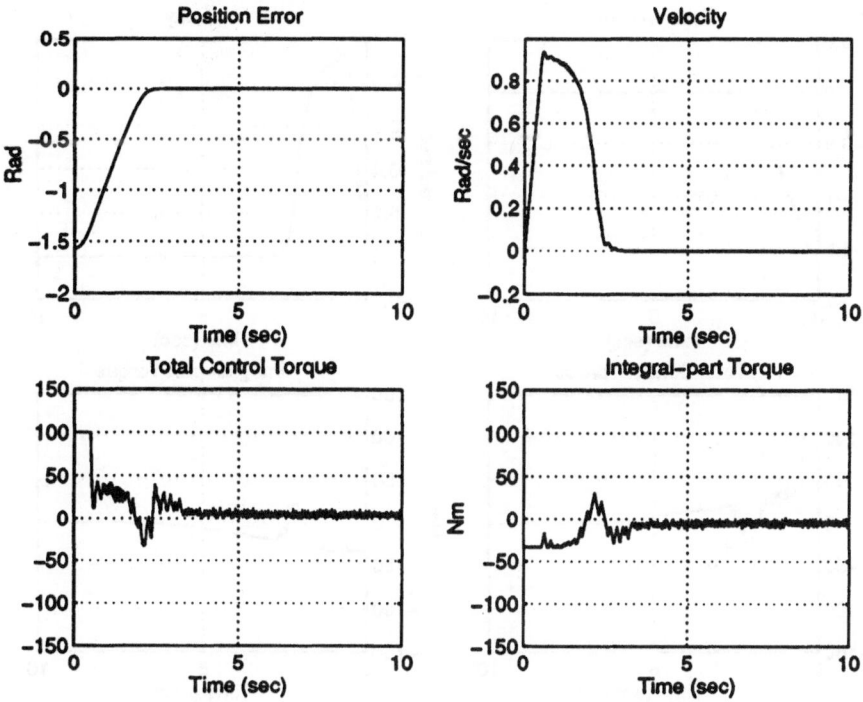

Figure 26: Experimental results obtained by using the switching integral control law (82).

obtained by using the following set of control parameters

$$\omega_0 = 10, \qquad \dot{q}_m = 1, \qquad K_0 = 1000,$$
$$K_s = 200, \qquad \gamma = 1/3, \qquad U_m = 100 \tag{83}$$

With this controller, the controlled system shows a smooth dynamic behaviour in the transient and a zero position error in the steady state condition. However, a small oscillation is present on the control variable.

A modified version of switching control law (82) has been tested in order to further reduce the chattering on the control action, as follows:

$$\begin{cases} u(t) = -K_0\sigma - r \\ \dot{r} = \begin{cases} K_s \, \text{sat}\,(h_s\sigma/K_s) & \text{if} \quad |r| \le \gamma U_m \\ -\rho r & \text{if} \quad |r| > \gamma U_m \end{cases} \end{cases} \tag{84}$$

In Fig. 27 are reported the experimental results obtained by using control law (84) with the same parameters values used in (83) and with

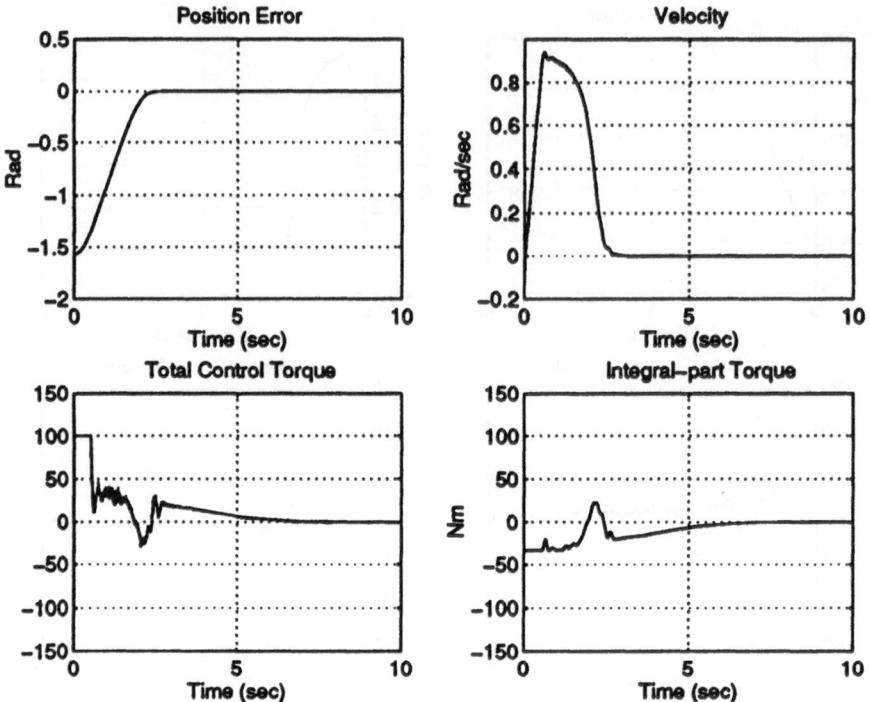

Figure 27: Experimental results obtained by using the saturated integral control law (84)

$h_s = 20000$. In fact, this controller provides the same smooth dynamic behavior obtained with the previous controller (82) but with the advantage that in this case the chattering on the control variable is no more present.

6 Conclusions

In the framework of the Variable Structure Control theory, a set of simple-to-implement control laws aiming to alleviate the chattering problem has been illustrated in a coordinated way. The underlying approach which is called Discontinuous Integral Control, originally presented for the continuous-time case, has been developed for the discrete-time case with a variety of modifications for dealing with known and unknown system parameters.

The experimental results related to some significant application cases here explicitly reported, and others obtained during the development of our long-term UBH project [18] [28], seem to confirm the effectiveness of the DIC technique with respect to both the expected performances and the simplicity of the technological implementation.

As regards possible methodological and applicative developments, one can mention the use of VSC and in particular of DIC paradigm in defining systematic design procedures of fuzzy logic controllers [25], [26]. Furthermore, the use of general classes of nonlinear sliding surfaces including the presence of inputs and their time-derivatives for chattering-free dynamic controller definition seems to be an interesting research stream [27]. Finally, as far as the application side is concerned one can note that, despite the growing number of successful application cases reported in the literature, the concepts of Variable Structure Control in general, and DIC in particular, need the development of easy-to-use CAD tools in order to become a popular design technique in control engineering.

Appendix A: Proof of Result 4

From (68) it follows that

$$\Delta\bar{e}(n) = \frac{J_e(n-1)}{J}e(n) + k\left[1 - \frac{J_e(n-1)}{J}\right]\mathrm{sgn}\, y(n-1)$$
$$-\frac{J_e(n-2)}{J}e(n-1) - k\left[1 - \frac{J_e(n-2)}{J}\right]\mathrm{sgn}\, y(n-2)$$

By using the sliding mode condition $\mathrm{sgn}\, y(n-2) = -\mathrm{sgn}\, y(n-1)$, the updating equation (71) transforms as follows

$$
\begin{aligned}
J_e(n) &= J_e(n-1) + q\Delta\bar{e}(n)\,\mathrm{sgn}\, y(n-1) \\
&= J_e(n-1) + 2kq\left[1 - \frac{J_e(n-1) + J_e(n-2)}{2J}\right] + \\
&\quad + q\left[\frac{J_e(n-1)}{J}e(n) - \frac{J_e(n-2)}{J}e(n-1)\right]\mathrm{sgn}\, y(n-1)
\end{aligned}
\tag{85}
$$

If we introduce the *inertia estimation error* $J_z(n) = J_e(n) - J$ as new state space variable substituting $J_e(n)$, equations (69) and (85) are transformed as follows

$$
\begin{cases}
e(n+1) = \dfrac{J_z(n-1)}{J}\, e(n) + \Delta\overline{\psi}(n+1) + k\,\dfrac{J_z(n-1)}{J}\,\text{sgn}\,y(n-1) \\[2mm]
J_z(n) = J_z(n-1) - q\dfrac{k}{J}\,[J_z(\dot n-1) + J_z(n-2)] \\[2mm]
\qquad + q\left[1 + \dfrac{J_z(n-1)}{J}\right] e(n)\,\text{sgn}\,y(n-1) \\[2mm]
\qquad - q\left[1 + \dfrac{J_z(n-2)}{J}\right] e(n-1)\,\text{sgn}\,y(n-1)
\end{cases}
\tag{86}
$$

Let us consider the case when the external disturbance $\psi(t)$ is constant: $\Delta\overline{\psi}(n+1) = 0$. If we introduce the new state space variables

$$
x_1(n) = \frac{e(n)}{J}, \qquad x_2(n) = \frac{J_z(n)}{J}, \qquad x_3(n) = \frac{J_z(n-1)}{J}
$$

and the auxiliary constant $k_j = k/J$, system (86) becomes

$$
\begin{cases}
x_1(n+1) = x_3\,[k_j\,\text{sgn}\,y(n-1) - x_1] \\[2mm]
x_2(n+1) = (1 - k_jq)x_2 - k_jqx_3 \\[1mm]
\qquad\qquad + q\,\{(1 + x_2)x_3\,[k_j\,\text{sgn}\,y(n-1) - x_1] \\[1mm]
\qquad\qquad - (1 + x_3)x_1\}\,\text{sgn}\,y(n) \\[2mm]
x_3(n+1) = x_2
\end{cases}
\tag{87}
$$

For the sake of brevity, in the right-hand side of equations (87) the state variables $x_1(n)$, $x_2(n)$ and $x_3(n)$ have been denoted as x_1, x_2 and x_3 respectively. System (87) can also be represented in the compact form $x(n+1) = F(x(n), n)$, where $x(n) = [x_1(n)\ x_2(n)\ x_3(n)]^T$ and $F(x(n), n)$ is a proper non linear function of the state $x(n)$. One can directly verify that $x_1 = 0$, $x_2 = 0$ and $x_3 = 0$ is an equilibrium point. Equivalently $e(n) = 0$ and $J_e(n) = 0$ is an equilibrium point for system (86). The stability of the discrete nonlinear system (87) in a neighborhood of this equilibrium point can be proved by using the first Lyapunov theorem. The Jacobian matrix of the system is

$$
A(x) = \frac{\partial F(x(n), n)}{\partial x(n)} =
\begin{bmatrix}
a_{11} & 0 & a_{13} \\
a_{21} & a_{22} & a_{23} \\
0 & 1 & 0
\end{bmatrix}
\qquad \text{where}
$$

$a_{11} = -x_3$

$a_{13} = k_j \operatorname{sgn} y(n-1) - x_1$

$a_{21} = -q\left[(1+x_2)x_3 + (1+x_3)\right] \operatorname{sgn} y(n)$

$a_{22} = (1 - k_j q) + q x_3 \left[k_j \operatorname{sgn} y(n-1) - x_1\right] \operatorname{sgn} y(n)$

$a_{23} = +q(1+x_2)\left[k_j \operatorname{sgn} y(n-1) - x_1\right] \operatorname{sgn} y(n) - k_j q - q x_1 \operatorname{sgn} y(n)$

The characteristic equation $|z\,I - A(0)| = 0$ of matrix $A(x)$ in the equilibrium point $x = 0$ is:

$$z^3 + (k_j q - 1)z^2 - k_j q\left[\operatorname{sgn} y(n-1)\operatorname{sgn} y(n) - 1\right] z + k_j q \operatorname{sgn} y(n-1)\operatorname{sgn} y(n) = 0$$

If we impose the sliding mode condition $\operatorname{sgn} y(n-1)\operatorname{sgn} y(n) = -1$ we obtain the time-constant equation

$$P(z) = z^3 + (k_j q - 1)z^2 + 2k_j q z - k_j q = 0 \qquad (88)$$

The roots of equation (88) are functions of the product $k_j q$. Equation

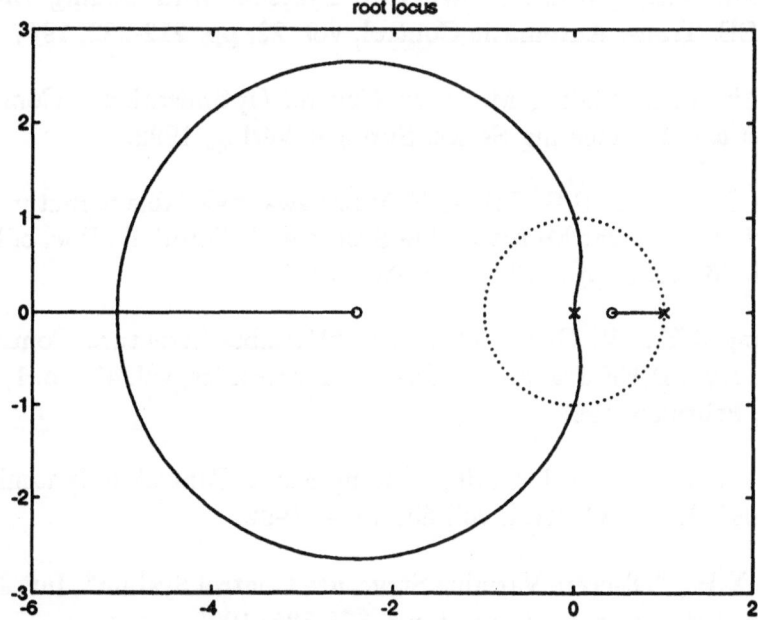

Figure 28: The global behavior of the root locus when $k_j q > 0$

(88) can be transformed as follows

$$1 + \frac{k_j q(z + 1 - \sqrt{2})(z + 1 + \sqrt{2})}{z^2(z - 1)} = 0$$

and the corresponding root locus when $k_j q$ varies from 0 to $+\infty$ results as reported in Fig. 28. By using the Jury's test one can easily find that the system is stable for $0 < k_j q < 0.5$, that is when $0 < q < J/(2k)$. From Fig. 28 it is evident that for $0 < k_j q < 0.5$ the three roots of equation (88) are inside the unit circle and then the first Lyapunov theorem assures that the nonlinear discrete system (87) is asymptotically stable in the equilibrium point $x = 0$. The stated Result is proved.   □

7   References

[1] Emelyanov S. V., "Variable Structure Control Systems", Moscow, Nauka, 1967.

[2] Utkin V.I., "Variable Structure Systems with Sliding Modes", IEEE, Trans. Automatic Control, vol. 22, pp. 212-222, 1977.

[3] Utkin V.I., "Sliding Modes in Control Optimization", Comunication and Engineering Series, Springer-Verlag, 1992.

[4] De Carlo R.A., S.H. Zak, G.P. Matthews, "Variable Structure Control of Nonlinear Multivariable Systems: A Tutorial", Poc. of IEEE, vol. 76, no. 3, pp. 212-232, March 1988.

[5] Hung J.Y., W. Gao, J.C. Hung, "Variable Structure Control: A Survey", IEEE Trans. on Industrial Electronics, vol. 40, no. 1, pp. 2-21, February 1993.

[6] Drakunov S.V., V.I. Utkin, "Sliding Mode Control in dynamic systems", Int. J. Control, vol. 55, no. 4, 1992.

[7] Yu X.H., "Discrete Variable Structure Control System", Int. J. Systems Science, vol. 24, no. 2, pp. 373-386, 1993.

[8] Sira-Ramírez H., "Nonlinear Discrete Variable Structure Systems in Quasi Sliding Mode", Int. Jour. of Control, vol. 54, no. 5, pp. 1171-1187, 1991.

[9] Zanasi R., "Discontinuous Control of Dynamic Systems with Applications in Robotics", Doctorate Thesis (tutor: C. Bonivento),

DEIS, University of Bologna, February 1992, (in Italian).

[10] Nersisian A., Zanasi R., "A Modified Variable Structure Control Algorithm for Stabilization of Uncertain Dynamical Systems", Int. J. of Robust and Nonlinear Control, vol. 3, pp. 199-209, 1993.

[11] Zanasi R., "Sliding Mode Using Discontinuous Control Algorithms of Integral Type", Int. Journal of Control, Special Issue on Sliding Mode Control, vol. 57, no. 5, pp. 1079-1099, 1993.

[12] Bonivento C., A. Nersisian, A. Tonielli, R. Zanasi, "A Cascade Structure For Robust Control Design", IEEE Trans. on Automatic Control, vol. 39, no. 4, pp. 846-849, April 1994.

[13] Bonivento C., R. Zanasi, "Discontinuous Integral Control", IEEE VSLT'94, Benevento, Italy, 1994.

[14] Bonivento C., M. Sandri, R. Zanasi, "Discrete Low-Chattering Variable Structure Controllers", European Control Conference 95, Roma, 5-8 Sept. 1995.

[15] Bonivento C., M. Sandri, R. Zanasi, "Discrete Variable Structure Integral Controllers", IFAC'96 - 13th World Congress, San Francisco, California, USA, June 30 - July 5 1996.

[16] Emelyanov S. V., "Binary Control Systems", MIR Publ., Moscow, 1988.

[17] Bonivento C., M. Sandri, R. Zanasi, "Binary Versus Sliding Mode: Simulation Experiments for Robotic Applications", IMACS-IFAC Int. Symp. on Math. and Intell. Models in Syst. Simulation, Brussels, Sept. 1990.

[18] Bonivento C., C. Melchiorri, R. Zanasi, "Robust Binary Control of Robotic Systems", invited paper in "Systems, Models and Feedback: Theory and Applications", A. Isidori and T.J. Tarn eds, Birkhauser, 1992.

[19] Bonivento C., A. Nersisian, A. Tonielli, R. Zanasi, "Robust Control Design Combining Binary and Variable Structure Techniques", NOLCOS'92, Bordeaux, France, June 1992.

[20] Sira-Ramirez H., Llanes-Santiago O., "Dynamical Discontinuous Feedback Strategies in the Regulation of Nonlinear Chemical Processes", IEEE Trans. on Control System Tech., vol. 2, no. 1, pp. 11-21, March 1994.

[21] Rossi C., A. Tonielli, "Robust Control of Permanent Magnet Motors: VSS Techniques lead to Simple Hardware Implementations", IEEE Trans. on Industrial Electronics, vol. 41, no. 4, August 1994.

[22] Bonivento C., C. Rossi, A. Tonielli, R. Zanasi, "Variable Structure Control of Electric Motors", Automazione e Strumentazione, 1996 (in Italian). In press.

[23] Zanasi R., R. Gorez and Y.L. Hsu, "Sliding Mode Control with Integral Action", Technical Report Nr.95.84 , Université Catholique de Louvain, Louvain-La-Neuve, Belgium, October 1995.

[24] Hsu Y.L., "Sliding Mode Control of Robot Manipulators", Ph.D Thesis, Université Catholique de Louvain, Louvain-La-Neuve, Belgium, 1995.

[25] Bonivento C., C. Fantuzzi, L. Martini, "Adaptive Fuzzy Logic Controller Syntesis Via Sliding Mode Approach", European Control Conference 95, Roma, 5-8 Sept. 1995.

[26] Bonivento C., C. Fantuzzi, "Showcase of a Fuzzy Logic Controller Design", IFAC'96 - 13th World Congress, San Francisco, California, USA, June 30 - July 5 1996.

[27] Sira-Ramirez H., "On the Dynamical Sliding Mode Control of Nonlinear Systems", Int. J. of Control, vol. 57, no. 5, pp. 1039-1061, 1993.

[28] C. Bonivento, C. Melchiorri, G. Vassura, R. Zanasi, "System Concepts and Control Techniques for the UBH Project", Second Int. Symp. on Measurement and Control in Robotics, ISMCR'92, Tsukuba Science City, Japan, Nov. 15-19, 1992.

Lecture Notes in Control and Information Sciences

Edited by M. Thoma

1992–1996 Published Titles:

Vol. 197: Henry, J.; Yvon, J.P. (Eds)
System Modelling and Optimization
975 pp approx. 1994 [3-540-19893-8]

Vol. 198: Winter, H.; Nüßer, H.-G. (Eds)
Advanced Technologies for Air Traffic Flow
Management
225 pp approx. 1994 [3-540-19895-4]

Vol. 199: Cohen, G.; Quadrat, J.-P. (Eds)
11th International Conference on
Analysis and Optimization of Systems –
Discrete Event Systems: Sophia-Antipolis,
June 15–16–17, 1994
648 pp. 1994 [3-540-19896-2]

Vol. 200: Yoshikawa, T.; Miyazaki, F. (Eds)
Experimental Robotics III: The 3rd
International Symposium, Kyoto, Japan,
October 28-30, 1993
624 pp. 1994 [3-540-19905-5]

Vol. 201: Kogan, J.
Robust Stability and Convexity
192 pp. 1994 [3-540-19919-5]

Vol. 202: Francis, B.A.; Tannenbaum, A.R.
(Eds)
Feedback Control, Nonlinear Systems,
and Complexity
288 pp. 1995 [3-540-19943-8]

Vol. 203: Popkov, Y.S.
Macrosystems Theory and its Applications:
Equilibrium Models
344 pp. 1995 [3-540-19955-1]

Vol. 204: Takahashi, S.; Takahara, Y.
Logical Approach to Systems Theory
192 pp. 1995 [3-540-19956-X]

Vol. 205: Kotta, U.
Inversion Method in the Discrete-time
Nonlinear Control Systems Synthesis
Problems
168 pp. 1995 [3-540-19966-7]

Vol. 206: Aganovic, Z.;.Gajic, Z.
Linear Optimal Control of Bilinear Systems
with Applications to Singular Perturbations
and Weak Coupling
133 pp. 1995 [3-540-19976-4]

Vol. 207: Gabasov, R.; Kirillova, F.M.;
Prischepova, S.V.
Optimal Feedback Control
224 pp. 1995 [3-540-19991-8]

Vol. 208: Khalil, H.K.; Chow, J.H.;
Ioannou, P.A. (Eds)
Proceedings of Workshop on Advances in
Control and its Applications
300 pp. 1995 [3-540-19993-4]

Vol. 209: Foias, C.; Özbay, H.;
Tannenbaum, A.
Robust Control of Infinite Dimensional
Systems: Frequency Domain Methods
230 pp. 1995 [3-540-19994-2]

Vol. 210: De Wilde, P.
Neural Network Models: An Analysis
164 pp. 1996 [3-540-19995-0]

Vol. 211: Gawronski, W.
Balanced Control of Flexible Structures
280 pp. 1996 [3-540-76017-2]

Vol. 212: Sanchez, A.
Formal Specification and Synthesis of
Procedural Controllers for Process Systems
248 pp. 1996 [3-540-76021-0]

Vol. 213: Patra, A.; Rao, G.P.
General Hybrid Orthogonal Functions and
their Applications in Systems and Control
144 pp. 1996 [3-540-76039-3]

Vol. 214: Yin, G.; Zhang, Q. (Eds)
Recent Advances in Control and
Optimization of Manufacturing Systems
240 pp. 1996 [3-540-76055-5]